Statistical Methods for Estimating Petroleum Resources

International Association for Mathematical Geology
STUDIES IN MATHEMATICAL GEOLOGY

1. William B. Size, Editor
Use and Abuse of Statistical Methods in the Earth Sciences

2. Lawrence J. Drew
Oil and Gas Forecasting: Reflections of a Petroleum Geologist

3. Ricardo A. Olea, Editor
Geostatistical Glossary and Multilingual Dictionary

4. Regina L. Hunter and C. John Mann, Editors
Techniques for Determining Probabilities of Geologic Events and Processes

5. John C. Davis and Ute Christina Herzfeld, Editors
Computers in Geology—25 Years of Progress

6. George Christakos
Modern Spatiotemporal Geostatistics

7. Vera Pawlowsky–Glahn and Ricardo Olea
Geostatistical Analysis of Compositional Data

8. P. J. Lee
Statistical Methods for Estimating Petroleum Resources

STATISTICAL METHODS FOR ESTIMATING PETROLEUM RESOURCES

P. J. Lee

OXFORD
UNIVERSITY PRESS
2008

Oxford University Press, Inc., publishes works that further
Oxford University's objective of excellence
in research, scholarship, and education.

Oxford New York
Auckland Cape Town Dar es Salaam Hong Kong Karachi
Kuala Lumpur Madrid Melbourne Mexico City Nairobi
New Delhi Shanghai Taipei Toronto

With offices in
Argentina Austria Brazil Chile Czech Republic France Greece
Guatemala Hungary Italy Japan Poland Portugal Singapore
South Korea Switzerland Thailand Turkey Ukraine Vietnam

Published by Oxford University Press, Inc.
198 Madison Avenue, New York, New York 10016

www.oup.com

Library of Congress Cataloging-in-Publication Data

Lee, P. J.
Statistical methods for estimating petroleum resources / P.J. Lee.
p. cm.
Includes bibliographical references and index.
ISBN 978-0-19-533190-5
1. Petroleum—Statistics. 2. Petroleum reserves—Statistics.
3. Petroleum industry and trade—Statistics. I. Title.
TN871.L374 2008
333.8′23015195—dc22 2007023993

9 8 7 6 5 4 3 2 1

Printed in the United States of America
on acid-free paper

Pei-Jen (P.J.) Lee
1934–1999

Oxford University Press mourns the loss of P.J. Lee, scholar, teacher, author, and friend. He completed work on this book just before his untimely death, and we gratefully dedicate it to him in recognition of his lifelong commitment to science.

Foreword to the Series

This series, Studies in Mathematical Geology (SMG), is issued under the auspices of the International Association for Mathematical Geology. It was established in 1984 by founding editor Richard B. McCammon to serve as an outlet for book-length contributions on topics of special interest to the geomathematical community and interdisciplinary branches that look to the Association for leadership in the application and use of mathematics in geoscience research and technology.

SMG no. 8 describes the underlying statistical concepts and methodology used by the PETRIMES system for petroleum resource assessment. Research on PETRIMES was initiated in 1979 at the Geological Survey of Canada by P. J. Lee, who acknowledged Prof. Gordon Kaufman's original discovery process model as his inspiration. The manuscript was written in 1999 during Prof. Lee's tenure at the National Cheng Kung University of Taiwan, shortly before his regrettably early, unexpected death. Originally intended as a text for graduate students, *Statistical Methods for Estimating Petroleum Resources* summarizes Prof. Lee's research on the topic. The PETRIMES system, widely used in the petroleum industry, continues to evolve.

Jo Anne DeGraffenreid, Editor
Baldwin City, Kansas, USA

Foreword

Oil and gas discovery process modeling bloomed during the late 1970s and early 1980s. P. J. Lee was a principal gardener. He nourished its development with passion. P. J. insisted that forecasts of undiscovered oil and gas in petroleum plays that he analyzed be based on sound geology and accurate modeling.

It is one thing to construct a model of oil and gas discovery in a petroleum play and publish a paper describing how to apply it to one or two example plays. However, a much larger and more difficult undertaking is the construction of a smoothly functioning system for the projection of future discoveries in each of a hundred plays with widely varying geological characteristics and discovery histories based on models of oil and gas discovery. P. J. was an intellectual spark plug who, with single-minded intensity, insisted on logical rigor, careful calibration, and constant improvement of just such a system: PETRIMES. He and his longtime colleagues Richard Procter and Paul Wang created this flagship of petroleum discovery systems. P. J. continued to modify and improve it until he left us—much too soon.

We are fortunate that he chose to write for us this account of discovery process modeling. It reflects his long and deep experience in applications of PETRIMES to petroleum plays throughout the world, and to plays in Alberta's Western Canada Sedimentary Basin in particular. In addition to being an invaluable record of research for a period of nearly 20 years, this manuscript is a benchmark for future research. It is required reading for the next generation of practitioners.

I have always been struck by the elegant fashion in which the Geological Survey of Canada presents its periodic summaries of Canadian oil and gas remaining to be discovered. These publications bear P. J.'s mark. If he were still with us, I know that he would be an enthusiastic participant in applying the new wave of computational methods washing over computer modeling and simulation.

I met P. J. soon after his arrival at the Institute of Sedimentary and Petroleum Geology. It was clear from the outset that he was beyond dedicated, driven perhaps, to *understand* and to *get it right.* Although I wish that I could have had more personal time with him, we maintained contact by letter and telephone as the years rolled by, and exchanged ideas and manuscripts. P. J. always graciously gave more credit to others than was due.

That segment of the oil and gas research community committed to projecting future discoveries of conventional oil and gas could do no better than to discover and adopt a P. J. Lee clone! We miss him personally and professionally.

Gordon M. Kaufman
Cambridge, Massachusetts, USA
April 22, 2007

Preface

It is with feelings of great honor and profound sadness that I introduce this work on behalf of my friend and colleague, Dr. Pei-Jen (P. J.) Lee, whose sudden death on November 1, 1999, deprived him of the opportunity. Traditionally, the author writes his own preface to provide readers with a window into the complicated writing process. I cannot open that window, but I hope to convey the object and scope of this monograph, P. J.'s last communication on an area of expertise that helped define his professional life.[1]

The scientific accomplishments of P. J. Lee are a huge source of pride for the Geological Survey of Canada. His body of work ranks among the Survey's most important contributions to petroleum geology, approaching the enunciation of the Anticlinal Paradigm by T. S. Hunt, the Survey's first geochemist. P. J. joined the Geological Survey of Canada in 1979. It was against the grand backdrop of the first and second "oil-price shocks" that he and his collaborators, most especially Paul Wang and Ping Tzeng, set out to provide Canada and the global geoscience community with improved tools to describe yet-to-be-found petroleum resources.

P. J. Lee dispassionately recounts these efforts in a review of petroleum assessments carried out by the Geological Survey of Canada (Lee, 1993d). That paper fails to capture the frantic atmosphere that prevailed as Canada set out to find its own "Prudhoe Bay" within its vast Arctic and oceanic frontiers. Neither does his paper capture the excitement that accompanied the internal availability of his new probabilistic methods at the Geological Survey in the early 1980s (Lee and Wang, 1983b, 1985). Especially exciting was the meeting with Gordon Kaufman and his team from the Massachusetts Institute of Technology (MIT), which provided encouragement for the continued development of the new methods. P. J. had a warm and wonderful smile; for days after Gordon departed, he literally beamed.

Between 1981 and 1990, Geological Survey colleagues came to work full of excitement and anticipating new ideas. The work of P. J. and his team reached a zenith with the development of the Petroleum Exploration and Resource Evaluation System (PETRIMES), which included both data management and resource assessment modules. The system was first described in 1989 in an unpublished Institute of Sedimentary and Petroleum Geology/Geological Survey of Canada user guide written by P. J. Lee and Ping Tzeng. In 1992, the federal government and a grateful nation acknowledged P. J.'s achievements with the presentation by the Governor General of the Commemorative Medal during the 125th anniversary of Canadian Confederation. In 1993, P. J. was honored with a gold medal at the first annual Celebration of Excellence in Information Management.

The impacts of the 1983 crude oil price collapse were calamitous for Canada's upstream petroleum industry. Hard times were a harbinger of the challenges that P. J. and his colleagues faced to gain acceptance for their methods and results. The first major application described undiscovered conventional crude oil in western Canada, and the results were criticized as wildly optimistic. But even we had been too conservative. In a decade, P. J. had to revise the assessment, once exploration found the median predicted ultimate undiscovered potential.

Acceptance came slowly; there were many seasons of "tough sledding," although natural gas resource assessments, beginning with the 1993 Devonian study, were more positively received. Even now, probabilistic assessments have not been as strongly embraced as deterministic petroleum system models. P. J., however, was a patient and passionate advocate of probabilistic methods. Thus it was with some reservation that he retired from the Geological Survey of Canada—proud of his accomplishments, but uncertain of his legacy.

P. J. Lee joined National Cheng Kung University of Taiwan in 1996. In Taiwan, P. J. was a tireless worker, embracing new challenges that included geotechnical and structural geology in a country with many geological hazards. Concurrently, he single-handedly revised the western Canada conventional crude oil assessment. He often worked to the point of exhaustion. The manuscript that resulted in this volume is but one product of his many efforts during the last three years of his life.

Particularly considering the expository material in the appendices, this monograph could well be used as a graduate-level text, as originally intended by the author. It serves also as a guide to PETRIMES, and as a reference that describes petroleum resource assessment topics in general. It summarizes assessment methods developed in Canada under P. J.'s leadership, supplemented by work untaken in Taiwan.

For P. J., the inspiration for all this work began with Prof. Gordon Kaufman's discovery process model. P. J. was extremely grateful for the contributions and support of his colleagues and assistants. The methods described herein were developed with Paul C. C. Wang in the 1980s. Ping Tzeng assisted with computer system developments, and Jui-Yuan Chang later improved the system with the addition of a Windows interface. Many professionals provided valuable input, including Gerry Reinson, Jim Barclay, Jim Podruski, Tony Hamblin, Doug Cant, Wendy Warters, Tim Bird, Jack McMillan, Paul Price, Peter Hannigan, Katrina Olsen–Heiss, Dick Procter, Gordon Taylor, Yuan-Chen Cheng, Ting-Fang Chou, Ming-Shan Chen, Ruozhe Qin, Yanmin Shi, Fritz Agterberg, David White, Doug Klemme, Roy Roadifer, Bill James, Charles Masters, Dick Mast, Don Singer, Gordon Dolton, Larry Drew, Jack Schuenemeyer, Richard Sinding-Larsen, Chang-Jo Chung, Bob Crovelli, Henry Coustau, David Forman, and Gordon Kaufman (all identified in the original draft of this preface). P. J. was grateful to the Geological Survey of Canada for many years of support, and he conveyed special thanks to Miss Hsiu Lun Hsu of Tainan, Taiwan, who prepared the original text and figures upon which this monograph is based. Manuscript preparation was supported by both the National Science Council of Taiwan and National Cheng Kung University.

What is P. J.'s legacy? In a world where global crude oil consumption exceeds 82.4 million barrels (MMbbls) per day and natural gas consumption exceeds 2750 billion cubic meters (BP, 2006), the demands for better resource management and exploration efficiency have sparked new interest in petroleum assessment.

The Geological Survey of Canada continues to use PETRIMES to inform key national policy decisions and to assist revitalized frontier exploration. In addition, probabilistic methods find a wider acceptance and impact. Play definitions resulting from the western Canada gas assessments of the 1990s remain the template for recent assessments of ultimate conventional petroleum potential. The potential "supply gap" identified as a result of many Canadian assessments motivated the Geological Survey of Canada and its partners to provide global scientific leadership in the realization of unconventional resources from natural gas hydrates.

But most important, P. J.'s work of developing newer and sharper tools for resource assessment that find a new audience in environmental, economic, and policy communities continues at the Geological Survey of Canada (Chen and Osadetz, 2006; Gao et al., 2000). These developments provide a legacy of which a considerate gentleman, devoted husband and father, and active community enthusiast would be proud.

Kirk Osadetz
Geological Survey of Canada, Calgary
April 24, 2007

Note

1. Citations in the Preface appear in the list of references.

Acknowledgments

The appearance of this monograph would have been impossible without the help of the author's many friends and colleagues. Prof. Gordon Kaufman, MIT/Sloan School of Management, Cambridge, Massachusetts (who also kindly provided the Foreword); U.S. Geological Survey petroleum geologist Dr. Larry Drew, Reston, Virginia; and at the Kansas Geological Survey, Prof. Dan Merriam (Syracuse University, New York/University of Wichita, Kansas) teamed up to start the ball rolling. Thus a nearly legible, almost complete photocopy of Prof. Pei-Jen "P. J." Lee's final manuscript eventually arrived on the doorstep of the International Association for Mathematical Geology monograph editor. SMG no. 8, *Statistical Methods for Estimating Petroleum Resources*, was about to be born—in the fullness of time.

Prof. John Davis, University of Kansas/Montanuniversität Leoben, Austria, obliged the monograph midwife by scanning the original text, cleaning up illustrations, and patiently answering many questions. He also corresponded with Prof. Lee's colleagues at the Geological Survey of Canada; National Cheng Kung University, Taiwan; and throughout the United States.

Dr. Kirk Osadetz, P. J.'s close friend and coworker at the Geological Survey of Canada in Calgary, prepared the Preface for the monograph. Drawn as it was, in part, from P. J.'s original draft, he modestly omitted his own name from the list of those Prof. Lee wished especially to thank; it is hereby reinstated. Dr. Richard Procter of Calgary, a personal friend and colleague of P. J., served as liaison between the Lee family and the SMG editor and publisher. Dr. Zhouheng Chen, Geological Survey of Canada, Ottawa, was able to supply fair copies of several illegible manuscript pages. Correspondence with colleagues in Taiwan resulted in recovery of a missing table, provided by P. J.'s doctoral student, Prof. Yuan-Chen Cheng.

All of us who worked to convey P. J. Lee's final manuscript to graduate students and members of the petroleum industry and geoscience community owe a debt of gratitude to Dr. Geoff Bohling of the Kansas Geological Survey, University of Kansas. It was his task to renovate myriad equations that succumbed to the vagaries of word processing, optical character reader technology, and brutal, repeated photocopying. Geoff did an excellent job. The revised manuscript for *Statistical Methods for Estimating Petroleum Resources* was reviewed by SMG associate editor Thomas A. Jones (Exxon Production Research/Rice University, Houston, Texas), who brought several typographical blunders to my attention. The remaining errors are mine.

Jo Anne DeGraffenreid, SMG Editor

Contents

Statistical Methods for Estimating Petroleum Resources

1

Introduction

> In order to reach the Truth, it is necessary, once in one's life, to put everything in doubt—so far as possible.
>
> —Descartes

Background

Petroleum resource evaluations have been performed by geologists, geophysicists, geochemists, engineers, and statisticians for many decades in an attempt to estimate resource potential in a given region. Because of differences in the geological and statistical methods used for assessment, and the amount and type of data available, resource evaluations often vary. Accounts of various methods have been compiled by Haun (1975), Grenon (1979), Masters (1985), Rice (1986), and Mast et al. (1989). In addition, Lee and Gill (1999) used the Michigan reef play data to evaluate the merits of the log-geometric method of the U.S. Geological Survey (USGS); the PETRIMES method developed by the Geological Survey of Canada (GSC); the Arps and Roberts method; Bickel, Nair, and Wang's nonparametric finite population method; Kaufman's anchored method; and the geo-anchored method of Chen and Sinding–Larson.

Information required for petroleum resource evaluation includes all available reservoir data and data derived from the drilling of exploratory and development wells. Other essential geological information comes from regional geological, geophysical, and geochemical studies, as well as from work carried out in analogous basins. Any comprehensive resource evaluation procedure must combine raw data with information acquired from regional analysis and comparative studies.

The Hydrocarbon Assessment System Processor (HASP) has been used to blend available exploration data with previously gathered information (Energy, Mines and Resources Canada, 1977; Roy, 1979). HASP expresses combinations of exploration data and expert judgment as probability distributions for specific population attributes (such as pool area, net pay, porosity). Since this procedure was first implemented, demands on evaluation capability have steadily increased as evaluation results were increasingly applied to economic analyses. Traditional methods could no longer meet the new demands. A probabilistic formulation for HASP became necessary and was established by Lee and Wang (1983b). This formulation led to the development of the Petroleum Exploration and Resource Evaluation System, PETRIMES (Lee, 1993a, c, d; Lee and Tzeng, 1993; Lee and Wang, 1983a, b, 1984, 1985, 1986, 1987, 1990). Since then, new capabilities and features have been added to the evaluation system (Lee, 1997, 1998). A Windows version was also created (Lee et al., 1999). The statistical concepts and procedures used by PETRIMES and other methods are the topics of this book.

Objectives

The objective of an assessment is to evaluate the total resource or potential of a given region. The term *resource* is defined as the quantity of hydrocarbons of discovered and undiscovered pools; *potential* is defined as an undiscovered quantity of hydrocarbons. However, results of petroleum resource evaluations are usually given as aggregated numbers representing total resources. Aggregated potential values are not specific enough to be used in economic, exploration, or development planning analyses because all these processes require a knowledge of the number and size of undiscovered pools. Consequently, the objectives of a resource assessment are to

- estimate the number of yet-to-be discovered pools
- estimate the sizes of the undiscovered pools
- estimate the reservoir characteristics of the undiscovered pools
- validate exploration concepts with known information
- estimate pool-size distributions and relate these distributions to geological plays

The information provided by this type of assessment can be applied to economic analyses.

An Outline of the Evaluation Procedure

In this book, the procedure for resource evaluation is as follows:

1. Estimate pool-size distribution using either (a) the discovery process models for mature plays, which use superpopulation or finite population concepts; or (b) the multiplication of probability distributions of geological random variables according to a pool-size equation for conceptual or immature plays. In contrast to the definition adopted by Schuenemeyer and Drew (1983) and Davis and Chang (1989) (they define a field-size distribution in terms of the number of fields or pools), in this book, a *pool-size distribution* is defined as a pool-size probability distribution in terms of in-place or recoverable volume. Furthermore, a *pool* is defined as a single reservoir entity, and a *field* is a group of pools located within a geographic area.
2. Identify geological factors of a play and estimate their marginal probabilities.
3. Derive number-of-pools distribution from the operation of exploration risk and the number-of-prospects distribution.
4. Estimate individual pool sizes from the number-of-pools distribution and the pool-size distribution of a play.
5. Obtain the play resource and/or potential distribution.

The evaluation procedure outlined here was developed primarily for assessing petroleum resources. However, evaluation of mercury deposits (Lee and Singer, 1994) demonstrates that if ore deposits are classified according to their origins as "plays," as in petroleum geology, PETRIMES can then be applied.

Scope

Chapter 2 explains the meaning and applications of geological and statistical models in petroleum resource evaluation. In chapters 3 and 4, the superpopulation and finite population models, and data from the Beaverhill Lake play (for which a discovery record is available) are used to illustrate the resource evaluation procedure. In Chapter 5, a frontier play is used to illustrate the conceptual play evaluation procedure.

Chapter 6 contains the information and procedures needed to undertake an assessment, including the interaction between the assessors and the system, and the feedback mechanisms required. In Chapter 7, an overview of other assessment methods is presented. Chapter 8 presents a summary and guideline for choosing methods. Appendices A, B, C, and D present the statistical treatments of the methods.

2

Evaluation Models

How quaint the ways of paradox
At common sense she gaily mocks

—W. S. Gilbert

Geological Models and Play Definitions

The initial step in the evaluation of any petroleum resource is the identification of an appropriate geological population that can be delineated through subsurface study or basin analysis. A geological population represents a natural population and possesses a group of pools and/or prospects sharing common petroleum habitats. A natural population can be a single sedimentation model, structural style, type of trapping mechanism or geometry, tectonic cycle, stratigraphic sequence, or any combination of these criteria. Reasons for adopting these criteria in the definition of a geological model are the following:

- The geological population will be defined clearly and its associated resource can readily be estimated.
- Geologists can adopt known play data for future comparative geological studies.
- Geological variables of a natural population can be described by probability distributions (e.g., the lognormal distribution).

Statistical concepts such as the superpopulation concept can be applied to geological models so that, for specific plays, an estimate of undiscovered pool sizes can be made.

Figure 2.1 illustrates various sedimentary environments (tidal flat, lagoon, beach, and patch reef) that can be used as geological models in resource evaluation. Each of these models has its own distinguishing characteristics of source, reservoir, trapping mechanism, burial and thermal history of source beds, and migration pathway. In resource evaluation, to ensure the integrity of statistical analysis, each of these should be treated as a separate, natural population. Therefore, the logical steps in describing a play are (1) identify a single sedimentation model and (2) examine subsequent geological processes.

Geological processes such as faulting, erosion, folding, diagenesis, biodegradation, thermal history of source rocks, and migration history might provide a basis for further subdivisions of the model.

In some cases, two or more populations might be considered mistakenly as a single population because of a lack of understanding of the subsurface geology. If the resulting mixed population were to have two or more modes in its distribution, this could have an impact on resource evaluation results.

As an example, let us look at the Devonian Leduc reef trend from the Western Canada Sedimentary Basin (Reinson et al., 1993), as displayed

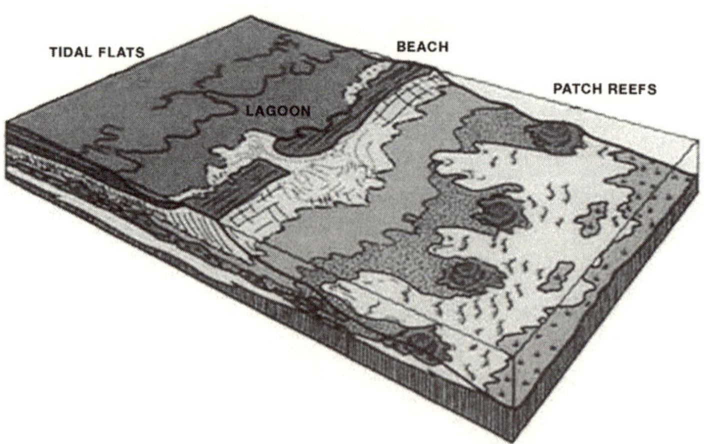

Figure 2.1. Examples of geological models: tidal flats, beach sand, patch reefs, and lagoon (after Wilson and Jordan, 1983). Each model may be defined as a basic unit for assessment.

in Figure 2.2. The setting includes the persistent Southern Alberta reef complex play, the Bashaw complex play, and the Ricinus–Meadowbrook isolated reef play. Reefs are deposited in a high-energy environment to form stromatoporoid rudstone, rooted to the carbonate platform of the underlying regressive hemicycle and persistent through the succeeding transgressive hemicycle. Traps on the carbonate shelf are controlled by transgressive–regressive hemicycles of a different order than the carbonate buildups in the persistent basinal facies belt (Wilson and Jordan, 1983). The traps along the Bashaw reef complex play exhibit a negative correlation between net pay and pool area, whereas the traps along the Ricinus–Meadowbrook chain exhibit a positive correlation. For petroleum evaluation, the three settings should be separated into three plays (Reinson et al., 1993).

Another example involves the Slave Point–Keg River succession (Reinson et al., 1993). In the northeastern part of British Columbia

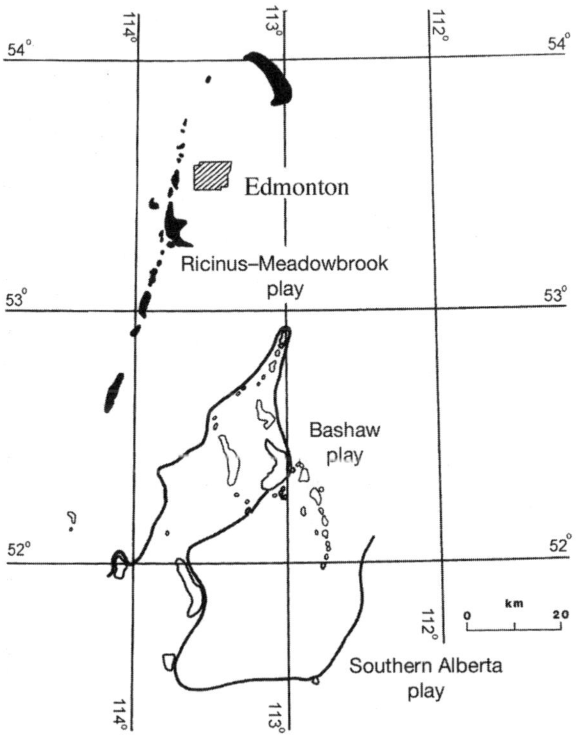

Figure 2.2. Leduc reefs (solid patches) in the Western Canada Sedimentary Basin divided into three plays: Ricinus–Meadowbrook, Bashaw, and Southern Alberta.

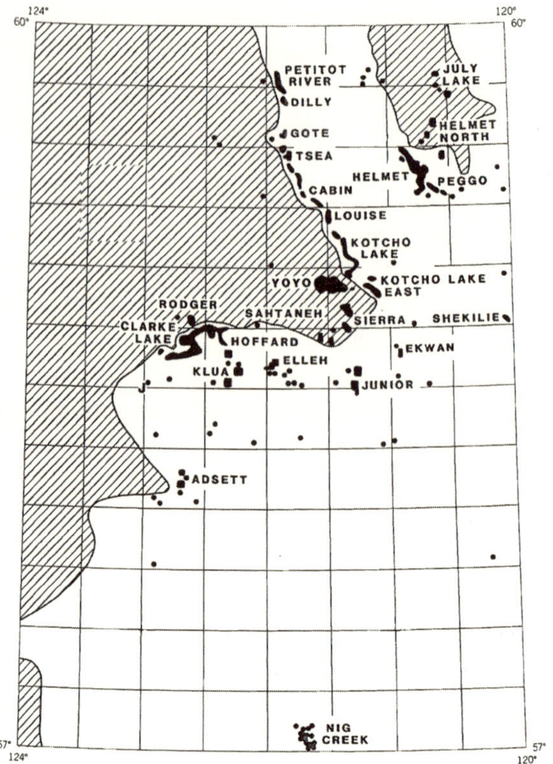

Figure 2.3. Slave Point–Keg River carbonate complex in northeastern British Columbia. Solid patches indicate reefs, identified as the Yoyo isolated reef play, Clarke Lake barrier reef play, and Adsett platform play.

(Fig. 2.3), the Middle Devonian Slave Point and Pine Point successions consist of two predominantly transgressive hemicycles separated by the Watt Mountain regression. A persistent Keg River–Sulphur Point–Slave Point carbonate barrier separates the evaporitic platform to the south and east from the Horn River basin to the north and west (Griffin, 1965a, b; Williams, 1984).

The lateral facies transition between these persistent carbonate and shale facies belts, referred to as the *facies front,* generally occupies a zone several kilometers wide and extends over a maximum stratigraphic interval of about 430 m. Prolific organic growth occurred at the front of the shelf, resulting in the formation of reef structures in places. Reservoirs along the rim of the platform are formed by barrier reefs, whereas isolated reefs form the reservoirs in the basin adjacent to the shelf. The Slave Point Formation, and probably the Sulphur Point and Pine Point formations, are dolomitized and diagenetically altered, resulting in enhanced reservoir development.

The Slave Point and Pine Point formations exhibit at least three types of reef population (i.e., isolated reef, barrier reef, and platform reef). The areal extent and net pay of these populations may be quite different. The effect of the geology on the accumulation of hydrocarbons might also differ. Consequently, the Slave Point and Pine Point formations in northeastern British Columbia are divided into three plays with respect to natural gas resource evaluation: the Yoyo isolated reef play, the Clarke Lake barrier reef play, and the Adsett platform play.

The point to be emphasized here is that the first step in any resource evaluation is to identify properly the geological populations that will serve as the framework for statistical evaluation. It is also important to remember that a geological population is merely a working hypothesis that should be revised or redefined as new information becomes available.

The next step in play identification is to define the minimum pool size within a play at the time the assessments are performed. After the minimum pool size is defined and the sample for the assessment has been collected, the statistical models can predict the pool sizes within the range represented by the sample with least uncertainty. Predictions made beyond the sample bear larger uncertainty than those within the sample range. This concept applies to all statistical estimation methods.

It must be emphasized that the geological population adopted here is a single and natural geological population—a play. On the other hand, Drew (1990) adopted an entire basin truncated by depth boundaries. The estimation method used for the pool-size distribution of a play and of a basin should not be the same. This is discussed in Chapter 4. What statistical and geological models entail and how they relate to one another are topics of discussion in the following sections.

Statistical Models

Random variables of a geological model (e.g., net pay or porosity) can be quantified with a set of possible attainable values. If we take the porosity values from a sandstone formation as an example, we find that some values occur more frequently than others. Thus, we can associate each porosity value with a real number or with a likelihood (the likelihood that the value will occur—a large number for a likely outcome and a small number for an unlikely one). In other words, all the porosity values of a formation will be associated with a probability that describes their likelihood of occurrence. All these values and their probabilities form a probability distribution.

We know the probability associated with each value, but we may not be able to explain the process that leads to the distribution. This class of physical phenomenon (a so-called *random* phenomenon), behaves "randomly" according to a probability distribution. Therefore, if a specimen from a given formation is sampled and we wish to predict the value of a particular variable for that sample, then the probability distribution of that variable must be known.

One of the steps in resource evaluation is to estimate the probability distributions of geological random variables. There are two types of distributions: discrete and continuous. Let us take, for example, a finite number of pools in a play. Certainly all pools constitute a finite population and will exhibit a discrete distribution (Fig. 2.4A). On the other hand, pool values can be thought of as coming from an infinite population that has a continuous probability distribution. This continuous probability distribution is called a *superpopulation distribution* (Fig. 2.4B).

In cases when we have a random sample or a very large sample set collected from a geological population, normal statistics can be used to construct a probability distribution of the population. For example, 406 porosity values have been obtained from the Lower Mannville Formation of the Western Canada Sedimentary Basin. This

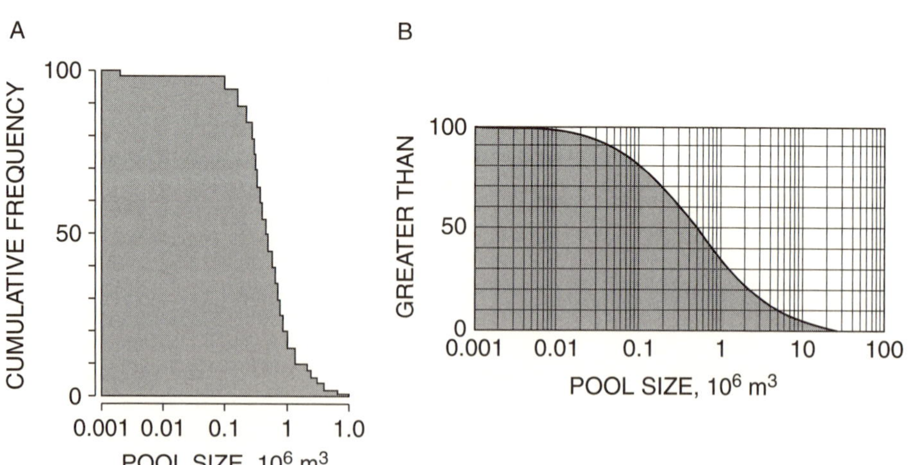

Figure 2.4. Examples of probability distributions. (A) Discrete distribution. (B) Continuous distribution.

sample set can be used to construct a histogram (Fig. 2.5A), a cumulative *greater-than* distribution (Fig. 2.5B), or a cumulative *less-than* distribution (Fig. 2.5C). These types of continuous distribution are considered to be superpopulations. The greater-than form is used to express probability distributions in petroleum resource evaluation. In reality, the sample sets of certain variables resulting from exploration are neither random nor large enough to represent the population. Therefore, specifics of the exploration discovery process are required if we are to estimate the mean and variance of the population.

Petroleum resource estimation procedures use the following statistical models:

- *The superpopulation and finite population models.* These models are needed to predict individual pool sizes in a population and to measure prediction uncertainties.
- *The discovery process model.* This model characterizes the discovery process and can be used to estimate the mean and variance of the population using data resulting from a selective discovery process.
- *The lognormal distribution model.* If a prior distribution such as a lognormal distribution is specified, then only the mean and variance of a population are required for the distribution to be estimated. The values for each percentile can be generated according to the lognormal distribution. On the other hand, if no prior distribution (nonparametric) is specified, then the values for each percentile must be estimated from the data.

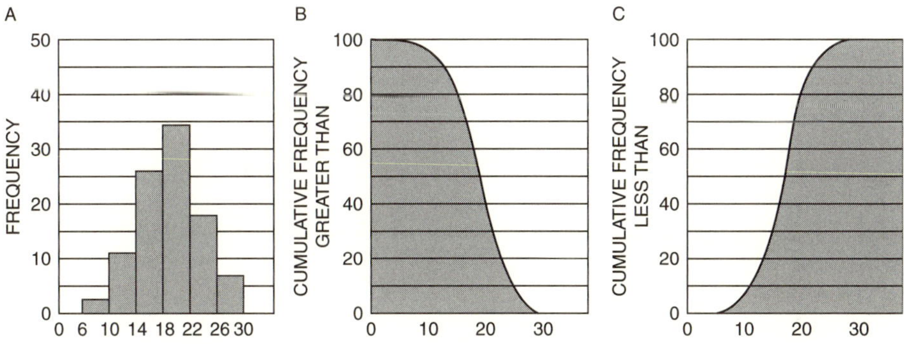

Figure 2.5. (A–C) Histogram (A), cumulative frequency greater-than plot (B), and cumulative frequency less-than plot (C) showing porosity distribution of the Mannville Formation, Western Canada Sedimentary Basin.

Concepts Used

Basic concepts used by PETRIMES are illustrated in figures 2.6 and 2.7. The upper right-hand corner of Figure 2.6 displays the facies distribution of a play containing pools and yet-to-be tested prospects. The discoveries from the play were plotted in terms of the discovery sequence (lower left-hand corner). Some questions and concerns that arise from examining the discovery sequence are as follows:

- How can these data be used to estimate the sizes of the undiscovered pools in this play?

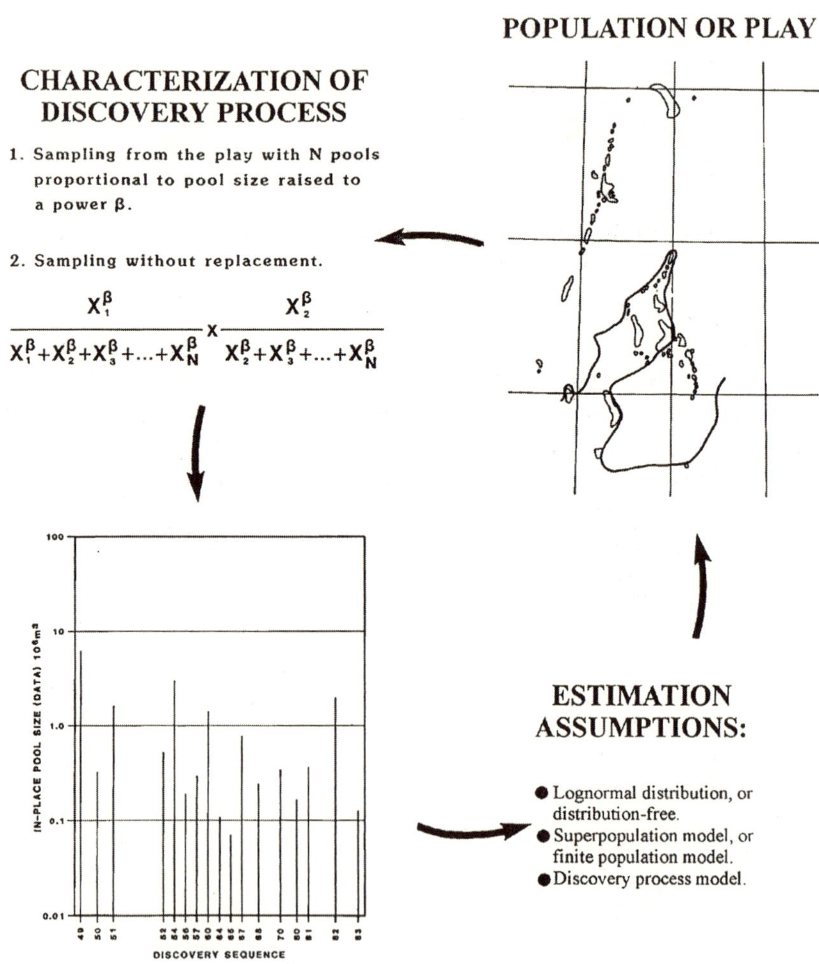

Figure 2.6. Sampling concept of the exploration discovery process.

- Can conventional statistical methods be used to predict undiscovered resources?
- If we adopt the usual method of computing the sample mean and variance for the population, the assumption is either that this is a random sample set from the population or that it is large

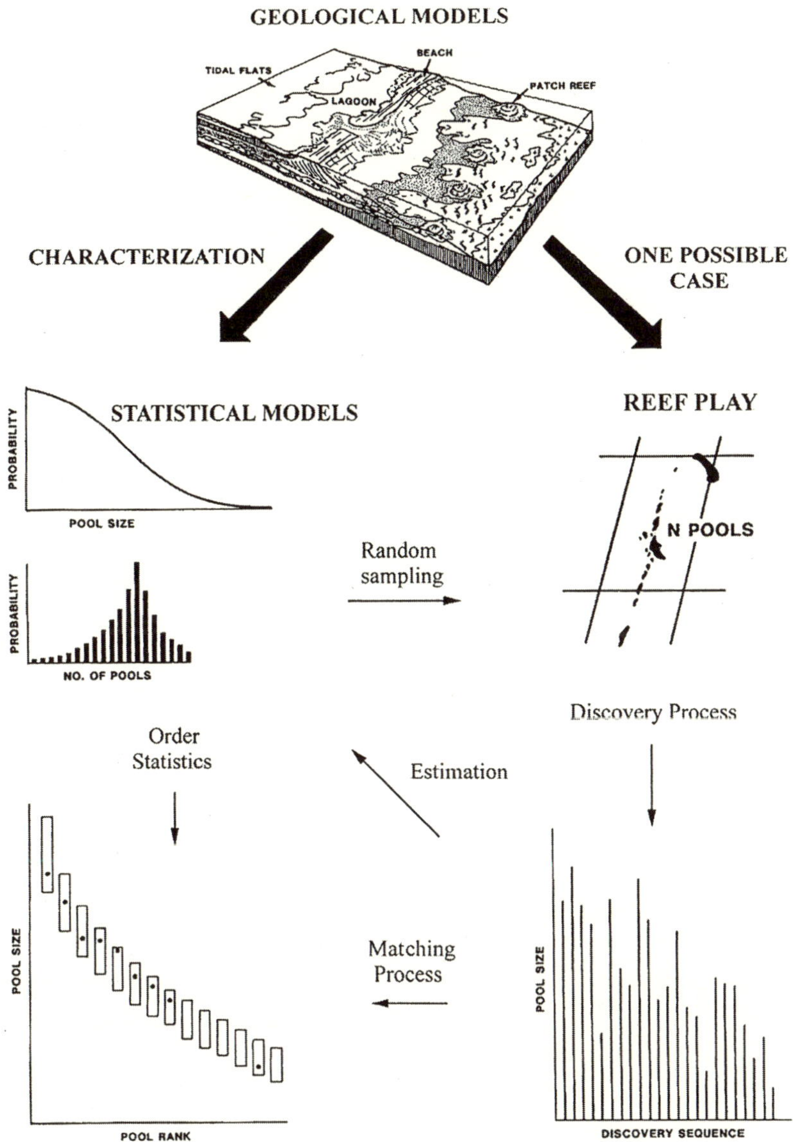

Figure 2.7. Statistical concepts used by PETRIMES.

enough to represent the population. In fact, neither of these assumptions is valid.

During the exploration–discovery process, large pools are normally discovered at an early stage. This implies that smaller pools remain to be discovered. Thus, the population mean would be overestimated by the sample mean obtained here, whereas the population variance would be underestimated by the sample variance. Therefore, we believe that the discovery process can be viewed as a sampling process whereby pool discovery probability is proportional to pool size and sampling without replacement.

Let us consider the patch reef model as an example of how statistical methods can be developed to evaluate a reef play. First, a reef model (Fig. 2.7, top) is defined as a collection of geologically analogous reef pools, and a reef play or population (upper right-hand corner of Fig. 2.7) contains some members of the reef model. In other words, a reef play consists of a finite number of reef pools, whereas a reef model contains an infinite number of reef pools with similar geological characters.

Second, a reef model can be described in terms of its geological random variables, such as pool size, pool area, net pay, porosity, and number of pools. The range of all possible values for each variable exhibits a continuous probability distribution because of the infinite number of reef pools, except that the number of pools has a discrete distribution expressed as an integer (Fig. 2.7, upper left-hand corner).

Third, for a specific play, the values of a variable are considered to be taken as a random sample from its probability distribution—in other words, they are independently derived from a common (or identical) distribution (written as *i.i.d.* in statistical literature). The following two statistical assumptions, which can be verified from basin analysis, are the following:

1. A play is defined as a single and natural population.
2. All pools are deposited under similar geological conditions.

Fourth, pool sizes obtained from discoveries of a play (lower right-hand corner of Fig. 2.7) can be used as a sample to estimate the two population distributions (continuous pool-size distribution and the discrete number-of-pools distribution).

In summary, two statistical assumptions are required: (1) all pools of a play have been deposited under similar geological conditions and (2) all pools within a specific play boundary form a single, natural geological population. Therefore, an adequate play definition would ensure that the subsequent statistical analyses are valid.

A play might contain many, few, or no discoveries at the time of evaluation. A play lacking discoveries (a conceptual play), or one containing few discoveries, is analyzed using the pool-size equation (see Chapter 5). If a play has sufficient discoveries (such as those shown in the lower right-hand corner of Fig. 2.7), there are two statistical approaches that can be applied to estimate the sizes of the remaining undiscovered pools.

The first approach, called the *superpopulation approach* (Baecher, 1979; Cassel et al., 1977; Cochran, 1939), is used to estimate the continuous pool-size distribution and the discrete number-of-pools distribution. The superpopulation approach views a play (the finite population) as one of the possible cases from the geological model (the infinite population or superpopulation), and has been described by Kaufman et al. (1975). The second approach is to estimate the play (upper right-hand corner of Fig. 2.7) without using the superpopulation concept. The play has a finite number of pools and a discrete pool-size distribution. This approach is called the *finite population approach.* Examples for adopting the finite population approach include the Arps and Roberts method (Arps and Roberts, 1958); Kaufman's anchored method (Kaufman, 1986); Bickel, Nair, and Wang's nonparametric finite population method (Bickel et al., 1992); and the geo-anchored method (Chen, 1993; Chen and Sinding–Larsen, 1992). In this book, both the superpopulation and the finite population approaches are discussed in chapters 3, 4, and 7.

When the superpopulation pool-size distribution and the number-of-pools distribution have been estimated, the individual pool sizes of the play can be estimated from order statistics, as shown in the lower left-hand corner of Figure 2.7. The boxes that express the estimation intervals can be matched with the current discoveries (shown in the lower right-hand corner). This matching process is one of several feedback mechanisms provided by PETRIMES that allow geological interpretations to be combined with statistical analysis.

In the following chapters, PETRIMES evaluation methods are validated using tested populations generated by known population parameters such as means and variances. The procedure for generating a finite number of pools from a superpopulation is described as follows:

- A hypothetical superpopulation with known mean and variance is assigned a probability distribution, such as the Pareto, lognormal, gamma, or Weibull distribution. This superpopulation can be considered a geological model.

- A random sample of size *N* is drawn from the superpopulation. This sample, which constitutes the finite number of pools, can be viewed as pools in a play.

The discovery process simulation was run with various exploration efficiencies (see hapter 3 for discussion) to generate different exploration time series, which could be used to verify the assessment results described in chapters 3 and 4.

The Nature of Geological Populations

Geological models have continuous population pool-size distributions that can be estimated from samples. Consequently, we must understand the nature of geological populations to choose probability distributions for them. In geological populations, properties such as outlier proneness and correlation of variables can be observed through analysis of two random variables. The Beaverhill Lake play and other oil plays from the Western Canada Sedimentary Basin are used in the following discussion to illustrate the nature of geological populations.

The Beaverhill Lake Play

Let us use the Late Devonian Beaverhill Lake play as an example for estimating a mature play. Transgression began with the deposition of the Slave Point carbonate on a broad shelf in northeastern British Columbia, northern Alberta, and the adjacent part of the Northwest Territories. A carbonate reef-front facies, similar to the underlying Elk Point reef carbonate, developed in British Columbia.

Continued transgression terminated the Slave Point carbonate platform, which was succeeded by basinal lithofacies of the overlying Waterways Formation in northern Alberta. However, in the Swan Hills region of north–central Alberta, a shallow-water platform, protected to the north by the emergent Peace River Arch and flanked to the southwest by the Western Alberta Ridge, provided a setting conducive to bank development and subsequent reef growth. Emergence of the reefs, followed by the rising water level during Beaverhill Lake deposition, terminated the growth of some Swan Hills reefs (Hemphill et al., 1968).

Subsurface study has revealed a sedimentation model in which the Slave Point carbonate platform pushed laterally into an open marine mudstone environment. Most of the discovered pools are situated along the platform margin or are adjacent to the platform (Fig. 2.8). Thus, the play contains traps related to organic buildups within the Beaverhill Lake carbonates of the Slave Point platform and deeper water equivalent sediments of the platform margin.

The play boundary was then delineated to reveal an oil play area that extends for more than 18,370 km^2. About 5230 wells have been drilled in this area, but only 844 wildcats have penetrated the Beaverhill Lake Group. From 1956 to 1988, 37 oil pools, as well as several gas pools, were discovered. These pools contain 949×10^6 m^3 (6 Bbbls) of in-place oil and $274{,}240 \times 10^6$ m^3 (9.7 Tcf) of in-place gas within the oil play boundary. The yield factors are 1.631 m^3/ha-m of recoverable oil (1.265 bbls/acre-ft.) and 395 m^3/ha-m of marketable gas (1730 cf/acre-ft.).

In addition to the 37 oil pools, 55 exploratory wells have shown oil in drill stem tests. It is assumed that these 55 wells are capable of producing for about 200 hours at the drill stem test rates. Therefore, their reserves were converted into in-place volumes using an average recovery factor of 0.10. These 55 pools were combined with the 37 discovered pools to form the discovery sequence to be used in the resource assessment (Fig. 2.9). Note that the Swan Hills A & B pool (Fig. 2.8) is separated into two

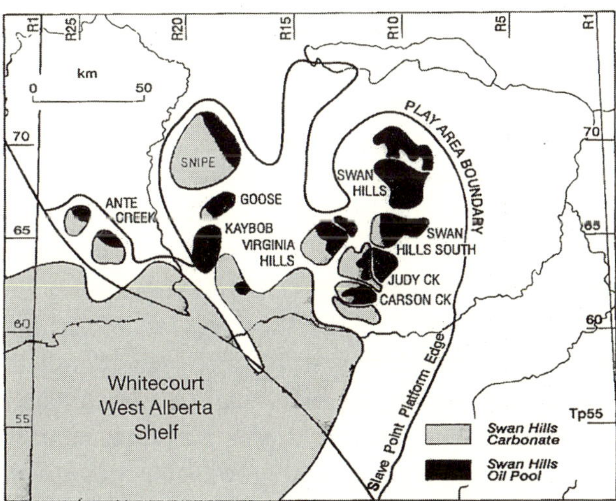

Figure 2.8. Facies map for Beaverhill Lake play, Western Canada Sedimentary Basin.

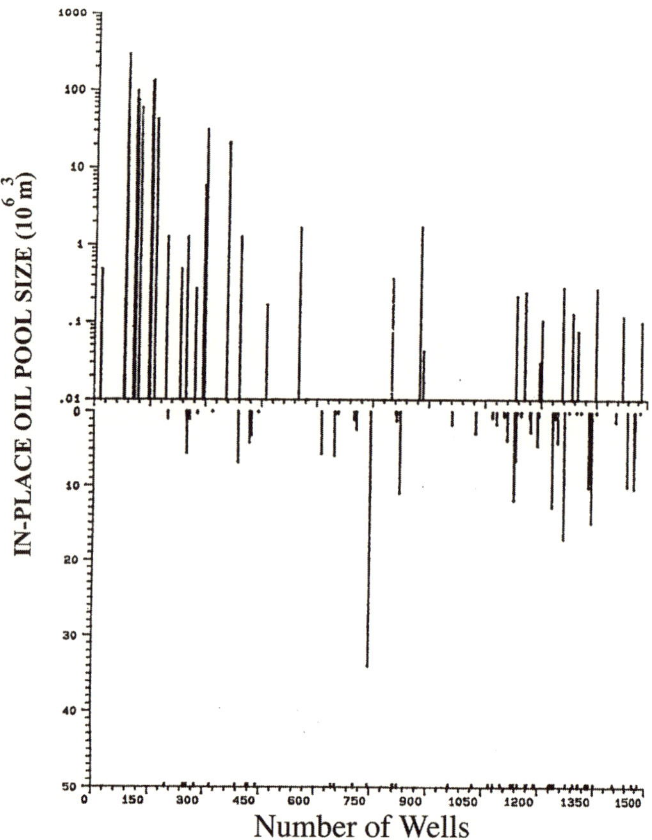

Figure 2.9. Exploration time series for Beaverhill Lake play. The upper half of the figure represents pools with commercial values; the lower half indicates oil recovered in drill stem tests.

pools with sizes 221×10^6 m^3 and 69×10^6 m^3. The upper half of Figure 2.9 displays the discovery sequence of all commercial pools. Gaps on the horizontal axis indicate failed exploratory wells. The lower half of Figure 2.9 displays results from drill stem tests. In this case, the minimum pool size defined for this play is 0.001 MMbbls in place. Resource evaluation can be performed on the discovery data of the upper half, or on the upper and lower halves together as an integral data set.

The reason for combining noncommercial pools with defined pools in an integrated discovery sequence is to obtain representation from the small pools. Additional statistical assumptions (such as a constant ratio between two adjacent size classes) are not required in this approach. On the other hand, the estimation of reserves from drill stem test results is time-consuming and requires reservoir engineering expertise.

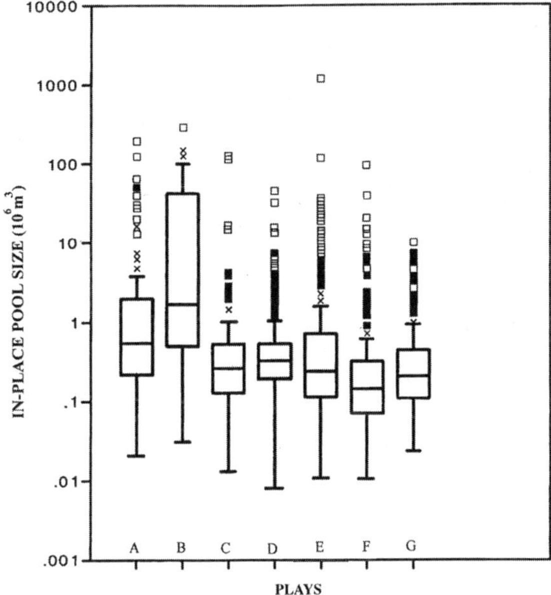

Figure 2.10. Box plots for in-place pool-size volume of several plays in the Western Canada Sedimentary Basin. A, Leduc reefs play; B, Beaverhill Lake play; C, Devonian sandstone play; D, Keg River reefs play; E, Cardium sandstone play; F, Viking sandstone play; G, Upper Mannville sandstone play.

Outliers

An *outlier* is a member of a population with either a relatively small or large value in comparison with other members of the same population. Outlier characteristics were described by Neyman and Scott (1971), who defined *outlier* and *outlier proneness* and demonstrated that distributions can be classified according to properties of their tails. If a population distribution has a long tail for the relatively large values (i.e., a large variance), then there is a higher probability of there being one or more outliers contained in the population. Both large and small outliers are observable in many geological populations, but only large outliers are discussed here.

Outliers can be recognized by plotting a variable on a box plot with a logarithmic scale. Box plots show where the median of a sample lies, and how the outliers relate to the median (Velleman and Hoaglin, 1981). For example, Figure 2.10 displays the box plots for the in-place pool size of several plays in the Western Canada Sedimentary Basin. In the box plot, the box covers the middle (50%) of the data. The horizontal bar within the box indicates the median of the sample, whereas the short vertical bar above the box covers the range occupied by three quarters of the data. The small squares and crosses outside the box indicate relatively large values. The largest one or two values in each sample are classified as outliers, the magnitudes of which are relative to the values of the sample. The difference between the largest and second largest pools of the Cardium sandstone play shown in Figure 2.10 is much greater than that of the other pools. Details for constructing a box plot are discussed in Lee et al. (1999).

Correlation between Random Variables

Correlation between geological random variables (such as pool area, net pay, recovery factor, reservoir pressure, and others) is also a common feature of geological populations. For example, the pool area and net pay variables of the Zama reef play of the Western Canada Sedimentary Basin exhibit a negative log–log association (Fig. 2.11A). In other words, as the log pool area value increases, the log net pay value decreases. In contrast, the pool area and net pay of the Beaverhill Lake play (Fig. 2.11B) show a positive log–log association: As the pool area value increases, the log net pay value increases. Correlation between

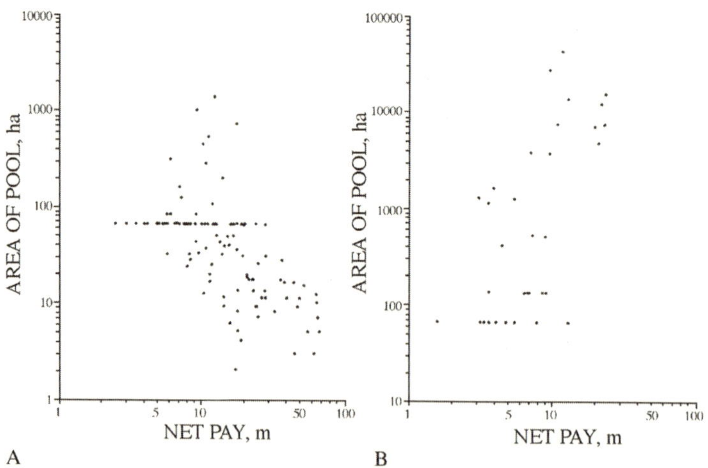

Figure 2.11. (A, B) Log–log associations for Western Canada Sedimentary Basin plays showing a negative correlation between pool area and average net pay variables for the Zama reef play (A), and a positive correlation between pool area and average net pay for the Beaverhill Lake play (B).

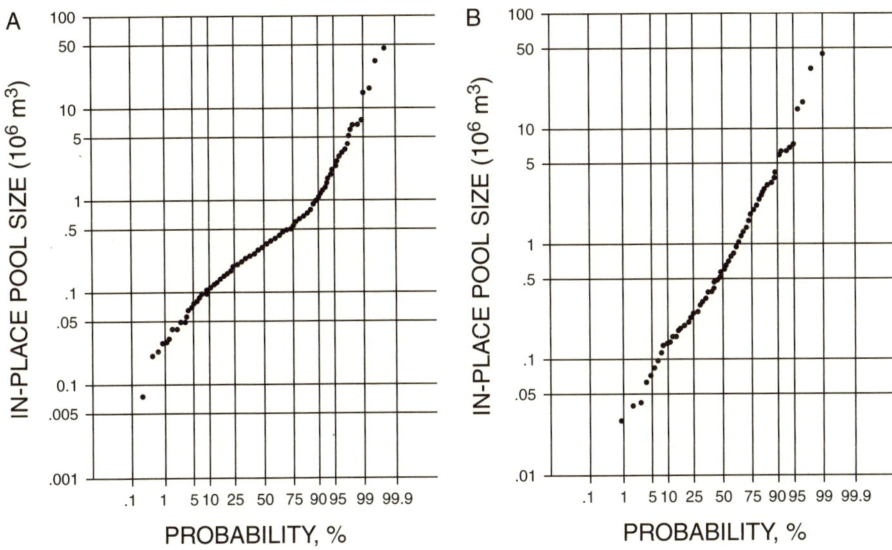

Figure 2.12. (A, B) Log probability plots for the Keg River reefs of the Black Creek basin (A) and the Rainbow basin (B).

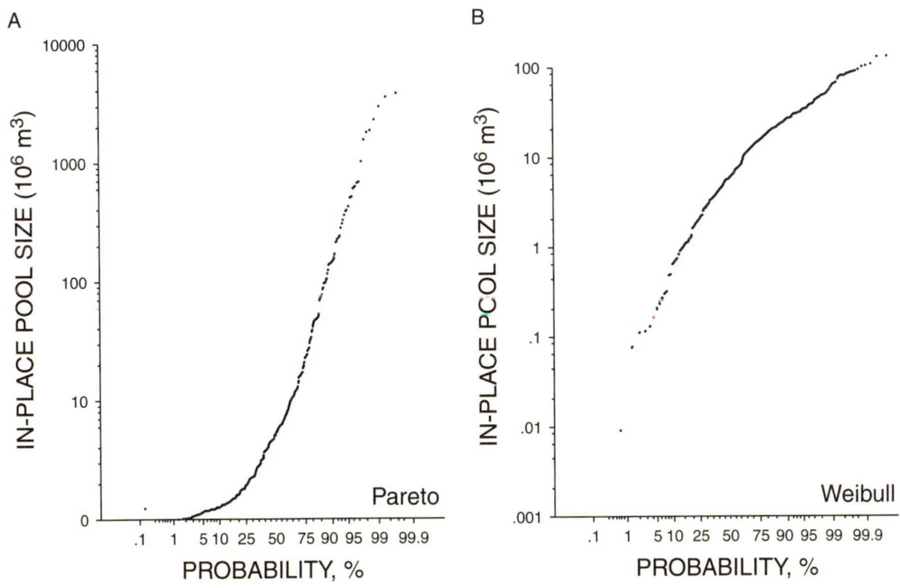

Figure 2.13. (A, B) Pareto population (A) and Weibull population (B) displayed on log probability graphs.

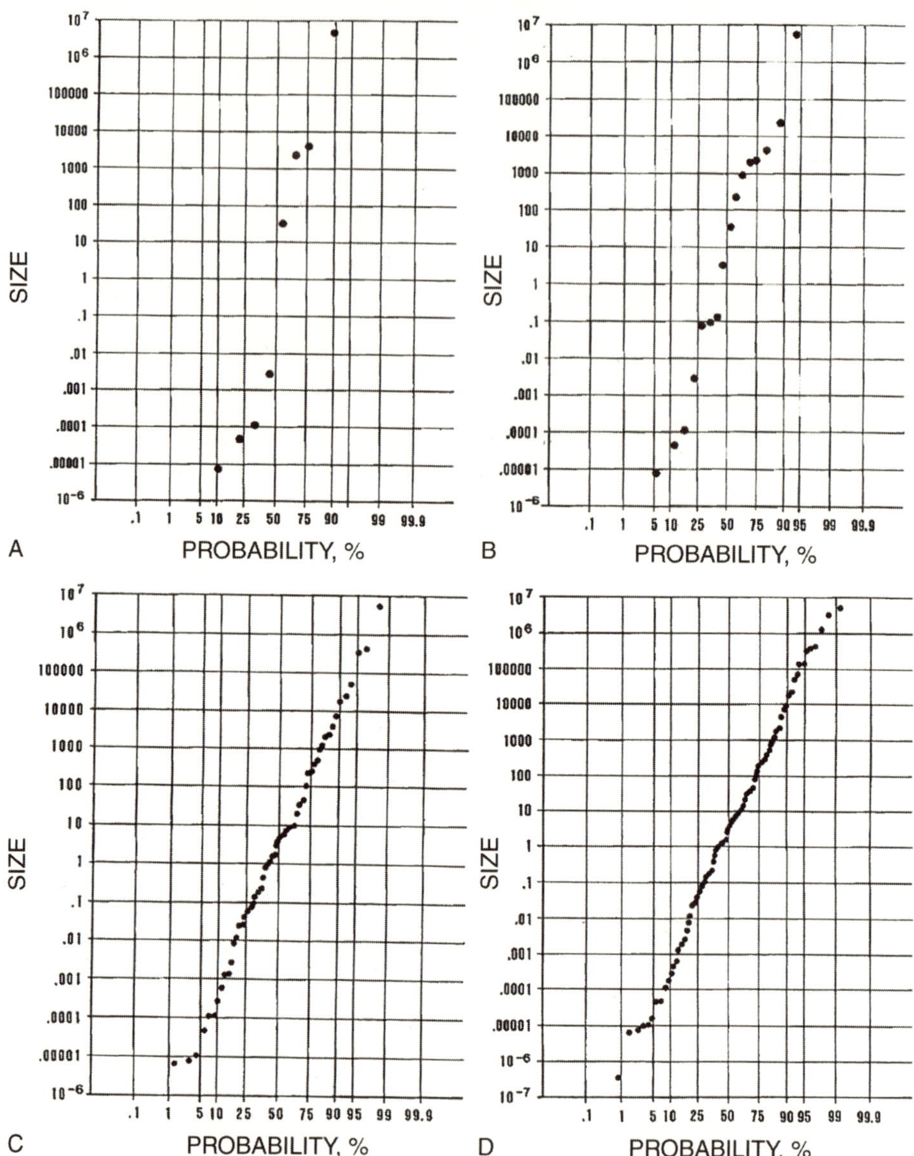

Figure 2.14. (A–D) Different sample sizes from a lognormal population displayed on log probability plots. Sample size: (A) $n = 8$, (B) $n = 16$, (C) $n = 64$, (D) $n = 128$.

variables is an important element to be considered in resource evaluation; otherwise, the mean and variance of a pool-size distribution may be over- or underestimated (see Chapter 5).

Mixed Populations

Figure 2.12A is a lognormal probability plot of all discovered Keg River reefs currently known from the Black Creek basin of the Western Canada Sedimentary Basin. The plot shown in Figure 2.12B displays the reefs from the Keg River shelf basin–Rainbow play, a subbasin within the Black Creek basin. Most of the data in Figure 2.12B follow a straight line, but the plot tends to be slightly convex upward. This convex-upward phenomenon may be the result of both dependent and biased sampling, because of the selective nature of the discovery process (i.e., large pools have higher probabilities of being discovered). Therefore, the nonlinearity in Figure 2.12A may be indicative of a mixed population.

The lack of linearity in the plot may be indicative of any one or all of the following circumstances:

1. The data set chosen is not from a lognormal population. Figure 2.13, for example, shows Pareto and Weibull data sets plotted on the log probability plot, which exhibits a serpentine pattern.
2. The data set was not chosen randomly (see Chapter 3).
3. There is more than one population in the data set (Fig. 2.12A).
4. The sample size is too small, as shown in Figure 2.14, which displays probability plots for a simulated lognormal distribution with different sample sizes. It is apparent that the plots become straighter when the sample size increases. The impact of mixed populations from lognormal, Pareto, and Weibull populations on the uncertainty of estimations will be discussed in Chapter 4.

From this overview of the nature of geological populations, we now move on in the next chapters to a discussion of how to apply these statistical models in petroleum resource evaluation.

3

Estimating Mature Plays

> A discovery process model is one built from assumptions that directly describe both physical features of the deposition of individual pools and fields and the fashion in which they are discovered.
>
> —Gordon M. Kaufman

A key objective in petroleum resource evaluation is to estimate oil and gas pool size (or field size) or oil and gas joint probability distributions for a particular population or play. The pool-size distribution, together with the number-of-pools distribution in a play can then be used to predict quantities such as the total remaining potential, the individual pool sizes, and the sizes of the largest undiscovered pools. These resource estimates provide the fundamental information upon which petroleum economic analyses and the planning of exploration strategies can be based.

The estimation of these types of pool-size distributions is a difficult task, however, because of the inherent sampling bias associated with exploration data. In many plays, larger pools tend to be discovered during the earlier phases of exploration. In addition, a combination of attributes, such as reservoir depth and distance to transportation center, often influences the order of discovery. Thus exploration data cannot be considered a random sample from the population. As stated by Drew et al. (1988), the form and specific parameters of the parent field-size distribution cannot be inferred with any confidence from the observed distribution. The biased nature of discovery data resulting from selective exploration decision making must be taken into account when making predictions about undiscovered oil and gas resources in a play. If this problem can be overcome, then the estimation of population mean, variance, and correlation among variables can be achieved. The objective of this chapter is to explain the characterization of the discovery process by statistical formulation.

To account for sampling bias, Kaufman et al. (1975) and Barouch and Kaufman (1977) used the successive sampling process of the superpopulation probabilistic model (discovery process model) to estimate the mean and variance of a given play. Here we shall discuss how to use superpopulation probabilistic models to estimate pool-size distribution.

The models to be discussed include the lognormal (LDSCV), nonparametric (NDSCV), lognormal/nonparametric–Poisson (BDSCV), and the bivariate lognormal, multivariate (MDSCV) discovery process methods. Their background, applications, and limitations will be illustrated by using play data sets from the Western Canada Sedimentary Basin as well as simulated populations. The steps for estimating undiscovered resources for a mature play involve (1) identifying a play, (2) compiling the data, (3) estimating pool-size distribution and number-of-pools distribution, (4) estimating pool-size-by-rank, (5) estimating play resource and play potential distribution, and (6) conducting feedback.

The superpopulation models do not require prior values for the total number-of-pools, population parameters, exploration efficiency, or truncation of large values. However, BDSCV requires a prior Poisson distribution for the number of pools and the lognormal pool-size distribution for estimating the posterior number-of-pools distribution. LDSCV requires a lognormal pool-size distribution, and MDSCV also requires a multivariate lognormal distribution for the reservoir parameters and a bivariate lognormal oil and gas pool-size distribution.

All available data are used to estimate population mean and variance, because an adequate estimate of population variance cannot be derived from truncated data. Furthermore, the procedure requires estimation of the population, rather than the fitting of a distribution to the discovery sequence.

The Superpopulation Model

Lognormal Discovery Process Model

In the superpopulation approach, the key step is to estimate the parameters of the underlying superpopulation distribution from samples obtained from exploration. Taking a lognormal distribution as an example, if the parameters—mean and variance—have been estimated, all the upper percentiles or the density of the distribution can then be generated.

We shall now discuss the principle of petroleum resource estimation from a statistical point of view. In cases in which the discovery data for a play come from a random sample or, alternatively, if all the discoveries have been made, the sample mean and variance adequately represent the population. However, in reality, discovery is influenced by many factors, including exploration techniques, drilling technology, acreage availability, and company objectives. Furthermore, geologists tend to test what is perceived to be the best or largest prospect, which might not be the largest pool of the play. Testing first for the best prospect tends to characterize the discovery process as a sampling procedure (as was indicated in Figure 2.9, which shows that discovered pool size gradually decreases with time). However, variations from that trend, or "waves," occur during the course of exploration. We are then faced with the question of how to use these types of biased samples to estimate the population. For the superpopulation model, a lognormal pool-size distribution is defined as

$$f_\theta(x) = \frac{1}{x\sigma\sqrt{2\pi}} \exp\left[-\frac{1}{2}\left(\frac{\ln x - \mu}{\sigma}\right)^2\right] \tag{3.1}$$

for $x > 0$, where $\theta = (\mu, \sigma^2)$ is the population parameter to be estimated. Examples of lognormal distribution shapes are presented in Figure 3.1. Here, μ is the mean of the population of logarithmic pool sizes and σ^2 is the variance of the population, n is the sample size (i.e., number of discoveries), and N is the total number of pools (discovered and undiscovered) in a play. The N value is also an unknown value to be estimated. A finite population was created from a random sample of size 300 ($N = 300$) drawn from the lognormal population with parameters $\mu = 3.0$ and $\sigma^2 = 5.0$. The histogram of the lognormal population (Fig. 3.2) exhibits a J-shaped distribution (the term *J-shaped* is used to describe a distribution monotonically increasing toward its left side) if an arithmetic scale is used for the horizontal axis. On the other hand, an almost symmetrical pattern results when a logarithmic scale is applied.

The estimation is based on the principle that the probability of discovering a pool is proportional to its size, and that a pool will not be discovered twice (Barouch and Kaufman, 1977; Kaufman, 1963; Kaufman et al., 1975). For the sake of simplicity, the concept of the

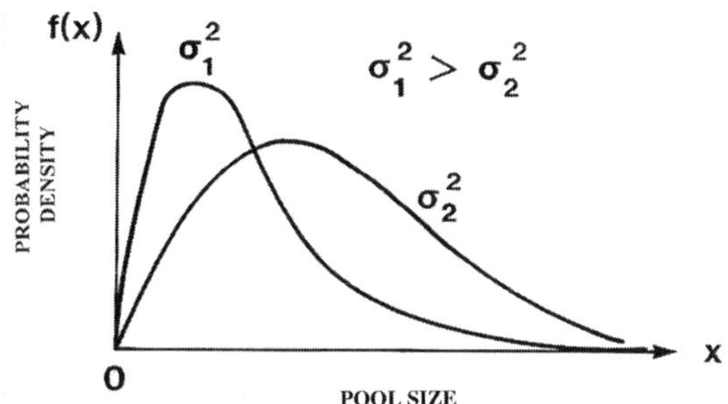

Figure 3.1. Examples of lognormal distributions:

$$f(x)=\frac{1}{x\sigma\sqrt{2\pi}}\exp\left[-\frac{1}{2}\left(\frac{\ln x-\mu}{\sigma}\right)^2\right],$$

where μ is the mean of the logarithmic transformed pool size, σ^2 is the variance of the logarithmic transformed pool size, and x is the pool size.

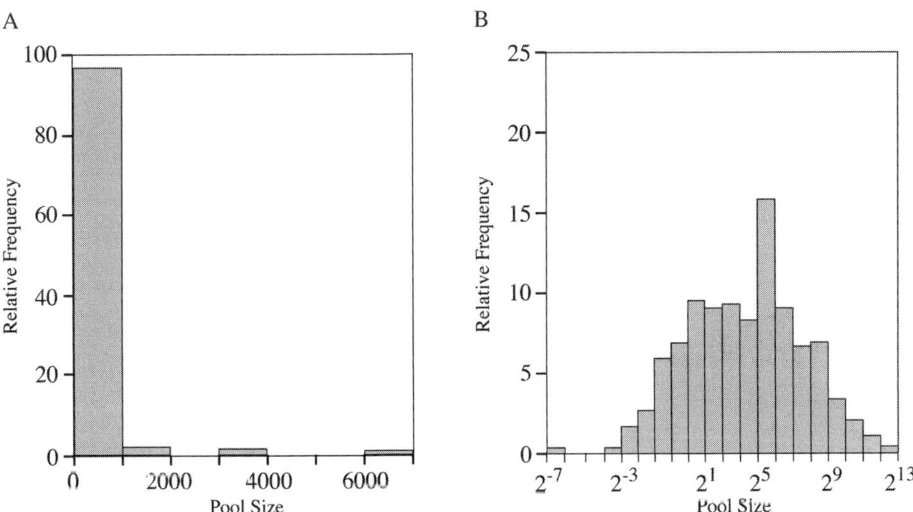

Figure 3.2. (A, B) Examples of a lognormal population. (A) Arithmetic scale. (B) Logarithmic scale.

discovery process model can be expressed as follows. The probability for pool j to be discovered is proportional to its size, x_j, as

$$P_j \propto \frac{X_j}{X_1 + \cdots + X_j + \cdots + X_N} \tag{3.2}$$

where $x_1, \ldots, x_N$ represents the pool size in the play, and N is total number of pools in the play.

Take the example of $N=3$ and $n=2$ to illustrate the discovery process model. Let the sizes of the three pools be $x_1=50$, $x_2=300$, and $x_3=100$ MMbbls. The probabilities for all possible discovery sequences are graphed in Figure 3.3, which indicates that the most likely sequence is (x_2, x_3, x_1), even though other sequences are also possible. This is the concept adopted by the discovery process model to characterize the exploration process. In other words, the probabilities for discovery of each pool of a play are set according to their volumes, and the probabilities for discovery of the remaining pools change as exploration continues. This concept allows us to formulate the discovery process likelihood function to be discussed in the following sections.

In Equation 3.2, the probability is completely proportional to pool size, but in reality pool size might be only one of many controlling factors. Thus, Equation 3.2 is generalized by adding an exponent to the equation as follows (Lee and Wang, 1985):

$$P_j \propto \frac{X_j^\beta}{X_1^\beta + \cdots + X_j^\beta + \cdots + X_N^\beta} \tag{3.3}$$

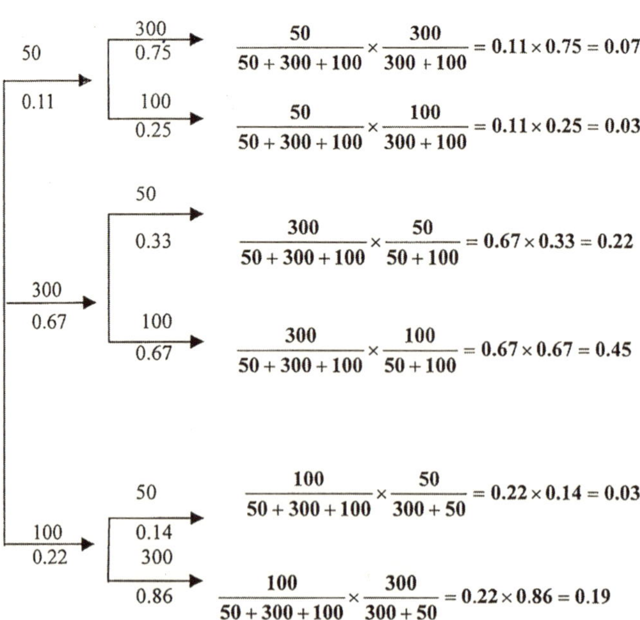

Figure 3.3. Examples of discovery sequence. $W_N = (100, 300, 50)$ and $N=3$, $n=2$.

where the β value ranges from negative to positive. The larger the β value, the greater the exploration efficiency will be. When $\beta=0$, the discovery process can be considered as a random sampling process. Therefore, the probability of observing $(x_1, \ldots, x_n)$, given Y_i, $i=1, \ldots, N$, is expressed as

$$P_j \langle X_1,\ldots,X_N \mid X_1,\ldots,X_N \rangle = \prod_{j=1}^{n} \frac{X_j^{\beta}}{b_j + Y_{n+1}^{\beta} + \cdots + Y_N^{\beta}} \tag{3.4}$$

where $b_j = x_j + \cdots + x_n$ (discovered pool sizes) and Y is equal to the undiscovered pool sizes.

The probability that the jth pool is deposited and discovered is the product of the following two probabilities: the probability of the deposition of a pool, j, with size, x_j, in the lognormal pool-size distribution, $f(x_j)$; and the probability of the pool j being discovered at a certain point in the sequence. Thus, the joint density function of all discovered pools can be shown as follows:

$$L(\theta) = \frac{N!}{(N-n)!} \prod_{j=1}^{n} f_\theta(X_j) \mathrm{E}_\theta \left[\prod_{j=1}^{n} \frac{X_j^{\beta}}{b_j + Y_{n+1}^{\beta} + \cdots + Y_N^{\beta}} \right] \tag{3.5}$$

where θ represents the distribution parameters (μ, σ^2), the factorial operation $N!/(N-n)!$ is the number of ordered samples of size n without replacement from a population of N pools, b_j is equal to $x_j + \cdots + x_n$ (discovered pools), and $y_{n+1}, \ldots, y_N$ is equal to the undiscovered pool sizes.

Quantity $L(\theta)$, which is the likelihood function of LDSCV, indicates the likelihood of a discovery sequence. What we attempt to do here is to reenact the exploration history. By doing so, we maximize the likelihood function by searching those values of μ, σ^2, and N for which the function $L(\theta)$ is maximized. The resultant $L(\theta)$ value is the maximized log-likelihood value. This procedure is called the *maximum-likelihood method* in statistics. The pool-size distribution $f_\theta(y)$ can be any probability distribution, but the lognormal family is applied here. In addition, the pool size variable can be replaced by any variable, such as pool area or net pay.

Equation 3.5 consists of two parts, f_θ and E[•]. The term f_θ represents the pool-size distribution, which results from tectonics, sedimentation, generation, migration, and accumulation of hydrocarbons, whereas E[•] represents the manner in which pools are discovered (Fig. 3.4).

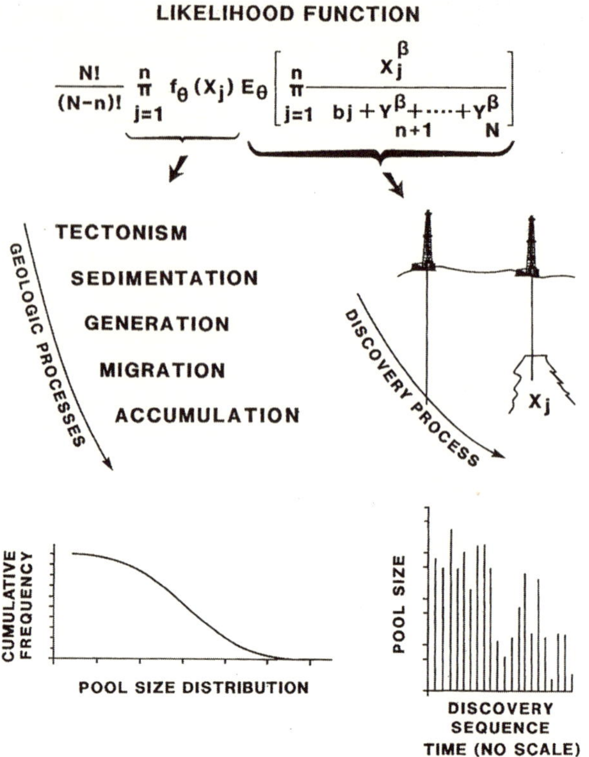

Figure 3.4. Diagram illustrating geological meanings of the lognormal discovery model.

If a data set were a random sample from its population, then E[•] and $N!/(N-n)!$ would be omitted from the likelihood function. That is to say, a random sample does not contain the information relating to the value of total number of pools in the play.

The contribution of the Kaufman model is that it not only characterizes the nature of the exploration process by setting the two basic assumptions, but it also expresses the likelihood function for the question: What is the probability of a pool with size x being deposited and also discovered at a certain point in the discovery sequence? Although it is true that this question is not of importance to explorationists, it does lead us to establish the likelihood function containing information on the superpopulation lognormal pool-size distribution and the total number of pools, N. Furthermore, the lognormal assumption has been used, but the model can handle all types of probability distributions. However, each probability distribution requires a specific numerical algorithm for its solution. The statistical treatment of the lognormal likelihood function is explained in Appendix A.

When Kaufman proposed this discovery process model, a numerical algorithm was used to solve the likelihood function ($\beta = 1$). Unfortunately, the algorithm is valid only when N is large (say, $N > 300$). The model has been criticized by statisticians and has been ignored by most petroleum assessment experts (who do not accept, or who are reluctant to accept, the principle of the discovery process model because petroleum geologists were not convinced by the example presented). Lee and Wang (1985) solved the likelihood function (Eq. 3.5) directly with an algorithm that can accommodate a wide range of values for total number of pools, N (the values tested ranged from 10 to more than 2000); number of discoveries, n (the values tested ranged from 9 to about 700); and β (ranging from –1 to 100). This algorithm requires intensive computation. Nevertheless, it provides reasonable predictions, as demonstrated by the populations tested. The successful solution of the likelihood function opens the possibility of using the discovery process models in petroleum resource assessments and improving their quality.

Nonparametric Discovery Process Model

A fundamental step in the probabilistic approach is to choose a prior probability distribution that the data obeys. So far, Kaufman (1963, 1965), Lee and Wang (1985, 1990), and Meisner and Demirmen (1981) have adopted the lognormal pool-size distribution to represent a superpopulation. The superpopulation framework, with its lognormal model, seems to be the most favored method, especially when the ratio of sample size (number of discovered pools) to total number of pools in the population is low. However, the choice of a prior probability distribution to describe pool-size distribution has been a controversial topic for the past several decades.

In the previous sections we demonstrated how to use the lognormal discovery process model (parametric)—LDSCV—to estimate pool-size distribution. We shall now discuss the use of a nonparametric model that does not benefit from a prior distribution.

A play contains N pools within the same underlying cumulative probability distribution F. If n pools are discovered randomly from the play, then the probability density for each pool is simply

$$p_i = \frac{1}{n} \tag{3.6}$$

Unfortunately, the n pools are not a random sample, but a biased sample from the play. Therefore, the statistical estimation of p_i requires use of LDSCV, as described earlier. On the other hand, with the discovery process model and the underlying empirical superpopulation distribution, p_i can also be estimated without making any assumptions about its shape, such as lognormal distribution.

As we have discussed for LDSCV, the likelihood function can adopt any probability distribution, such as a Weibull or Pareto distribution. Each distribution, however, would require a specific numerical algorithm to solve the likelihood function. Consequently, as a logical extension of the lognormal model, the birth of the nonparametric discovery process (NDSCV) ensued. It is used in the following ways:

- To estimate the empirical pool-size distribution and N nonparametrically
- To provide estimates of p_i to validate distributional assumptions
- To act as a validation tool for LDSCV

The statistical treatment of NDSCV is explained in Appendix B.

Estimating Pool-Size Distribution for the Beaverhill Lake Play

The in-place oil volumes and their discovery dates for the Beaverhill Lake play data set (shown in Fig. 2.9) were entered into LDSCV and NDSCV. The number of discoveries (sample size) equals 92. This data set includes commercial as well as noncommercial pools, with the smallest pool size equal to 0.001 MMbbls (1000 bbls).

Table 3.1, column 1, lists all the N values. For each N value, the values of μ, σ^2, β, and the log likelihood were estimated by LDSCV (columns 2 to 5) and by NDSCV (columns 6 to 9). The curve of log L versus N derived by both models increases rapidly (Fig. 3.5), but when $N > 400$, both curves increase slowly. On the other hand, if we examine the estimates from $N = 400$ to 500, we can visualize that by increasing the value of N, the number of small pools increases rapidly, whereas the number of pools for the midsize classes increases slowly. The point estimates for μ and σ^2 derived from both models when $N = 400$ (Table 3.1) are used in the matching process. From the estimated μ and σ^2, the corresponding lognormal and/or empirical pool-size probability distribution can be generated.

Figure 3.6 displays the estimated pool-size distributions derived from LDSCV (Fig. 3.6, line A, $\hat{\beta} = 0.4$) and NDSCV (Fig. 3.6, line B,

Table 3.1. The Log-Likelihood Value and Its Corresponding Estimated Values for the Beaverhill Lake Play

N	*LDSCV*				*NDSCV*			
	$\hat{\mu}$	$\hat{\sigma}^2$	$\hat{\beta}$	*Log L*	$\hat{\mu}$	$\hat{\sigma}^2$	$\hat{\beta}$	*Log L*
100	–3.887	17.53	0.2	–170.782	–3.79	16.29	0.2	–732.5
110	–4.297	18.67	0.2	–168.754	–4.10	16.23	0.2	–730.5
120	–4.745	19.940	0.3	–167.221	–4.45	16.32	0.3	–728.5
130	–5.009	20.90	0.3	–165.786	–4.69	16.08	0.3	–727.0
140	–5.419	21.73	0.3	–164.915	–4.89	15.78	0.3	–726.1
150	–5.711	22.48	0.3	–164.880	–5.07	15.46	0.3	–725.6
160	–5.978	23.15	0.3	–164.060	–5.22	15.13	0.3	–725.2
170	–6.226	23.76	0.3	–163.809	–5.46	14.92	0.4	–724.9
180	–6.455	24.31	0.3	–163.805	–5.59	14.60	0.4	–724.2
190	–6.670	24.83	0.3	–163.805	–5.71	14.28	0.4	–723.5
200	–6.961	25.45	0.36	–162.645	–5.82	13.98	0.4	–723.0
210	–7.169	25.94	0.36	–162.444	–5.92	13.68	0.4	–722.6
220	–7.366	26.40	0.37	–162.269	–6.01	13.40	0.4	–722.3
230	–7.554	26.83	0.37	–162.115	–6.093	13.12	0.4	–722.1
240	–7.734	27.25	0.38	–161.978	–6.17	12.86	0.4	–721.9
250	–7.906	27.64	0.38	–161.856	–6.24	12.61	0.4	–721.7
260	–8.072	28.03	0.39	–161.746	–6.30	12.37	0.4	–721.6
270	–8.230	28.39	0.39	–161.647	–6.36	12.15	0.4	–721.5
280	–8.38	28.74	0.39	–161.557	–6.42	11.93	0.4	–721.4
290	–8.53	29.08	0.40	–161.476	–6.47	11.73	0.4	–721.4
300	–8.67	29.40	0.40	–161.401	–6.52	11.53	0.4	–721.4
310	–8.81	29.71	0.40	–161.333	–6.56	11.34	0.4	–721.4
320	–8.94	30.01	0.41	–161.270	–6.70	11.03	0.5	–721.3
330	–9.07	30.30	0.41	–161.213	–6.75	10.84	0.5	–721.1
340	–9.20	30.59	0.41	–161.159	–6.79	10.67	0.5	–720.8
350	–9.327	30.86	0.41	–161.109	–6.83	10.50	0.5	–720.6
360	–9.44	31.13	0.42	–161.063	–6.87	10.34	0.5	–720.4
370	–9.55	31.38	0.42	–161.020	–6.90	10.19	0.5	–720.3
380	–9.66	31.64	0.42	–160.980	–6.94	10.04	0.5	–720.1
390	9.77	31.88	0.42	–160.943	–6.97	9.90	0.5	–720.0
400	–9.87	32.12	0.42	–160.908	–7.00	9.76	0.5	–719.8
410	–9.98	32.35	0.43	–160.875	–7.03	9.63	0.5	–719.7
420	–10.08	32.58	0.43	–160.844	–7.06	9.50	0.5	–719.6
430	–10.19	33.80	0.43	–160.825	–7.09	9.38	0.5	–719.5
440	–10.27	33.02	0.43	–160.788	–7.11	9.26	0.5	–719.4
450	–10.36	33.23	0.43	–160.762	–7.14	9.15	0.5	–719.3
460	–10.46	33.43	0.44	–160.737	–7.16	9.03	0.5	–719.3
470	–10.55	33.64	0.44	–160.714	–7.18	8.93	0.5	–719.2
480	–10.63	33.84	0.44	–160.692	–7.21	8.83	0.5	–719.1
490	–10.72	34.03	0.44	–160.672	–7.23	8.73	0.5	–719.1
500	–10.80	34.22	0.44	–160.652	–7.25	8.63	0.5	–719.0

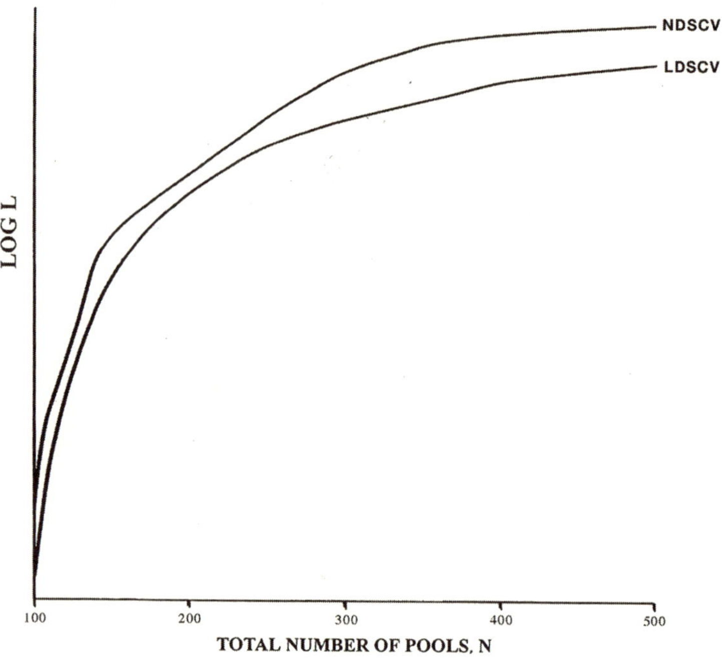

Figure 3.5. Diagram showing relationships between log-likelihood value versus the N value for the Beaverhill Lake play. LDSCV derived by the lognormal discovery process method. NDSCV derived by the nonparametric discovery process method.

$\hat{\beta}=0.5$) when $N=400$. The sample pool-size distribution (Fig. 3.6, line C) is shown in the same figure for comparison. It is evident that the mean and variance of the superpopulation pool-size distribution are over- and underestimated, respectively, if the pools are assumed to be randomly discovered (i.e., random sampling).

Lognormal/Nonparametric–Poisson Discovery Process Model

Previous Work

With the superpopulation concept, one can use a discrete probability distribution to express the number-of-pools distribution (Lee and Wang, 1990). The total number of pools that exists in a population is a value obtained from its superpopulation distribution. This section presents a statistical method referred to as the *Bayesian lognormal/nonparametric–Poisson discovery process model* (BDSCV) that is used to estimate the superpopulation discrete number-of-pools distribution

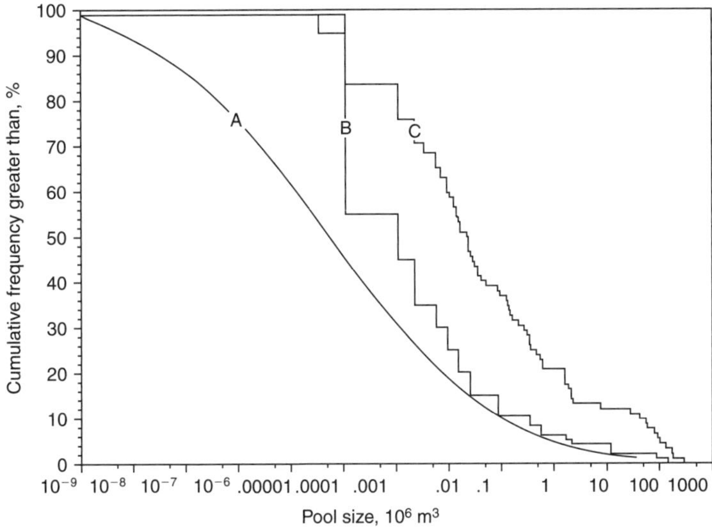

Figure 3.6. Pool-size distribution for the Beaverhill Lake play. Line A is derived by LDSCV, line B is derived by NDSCV, and line C is derived by random sampling.

when the discovery sequence is available. BDSCV provides a probability statement about the N value and a probability measure for each undiscovered individual pool size.

There are four methods for estimating the number of pools or the number-of-pools distribution for a play. The first method is as follows: Geophysicists and geologists obtain the number of prospects from structural contour maps based on seismic and geological information. The information about the number of prospects is used to construct a number-of-prospects distribution (Lee and Wang, 1990). The number-of-prospects distribution and the exploration risk (Lee et al., 1989) are used to derive the number-of-pools distribution (Lee, 1993d; Lee and Wang, 1983b).

The second method includes the maximum-likelihood methods of LDSCV and NDSCV, as discussed in the previous sections. However, both LDSCV and NDSCV can only provide point estimates about the value of N. Furthermore, the log-likelihood functions of these two methods occasionally show a flat profile about the value of N and yield a range of N without probability measures.

The third method (Gordon, 1983) is as follows: A successive sample from a finite population is divided into two parts to approximate the unknown inclusion probabilities. An estimate of N is obtained by an approximate Horvitz–Thompson-type estimator. This procedure requires solving a pair of symmetrical transcendental equations. Barouch et al. (1985) proposed an alternate pair of asymmetrical transcendental equations to solve the problem.

The fourth method postulates that N also has a superpopulation probability function, $P(\bullet|\gamma)$, indexed by a vector of parameters, γ, that is independent of the variate, the pool sizes x. The posterior distribution of γ is then used to make inferences about N. Here, the observations consist of x_n and $N > n$. The probability function, $P(N|\gamma)$, may be interpreted as a model describing a random mechanism of how N is generated, or it might be considered as a prior distribution in an empirical Bayesian context. Wang and Nair (1988) presented a lognormal case, which was extended as a generalized procedure (Lee, 1997).

The BDSCV Model

Four statistical assumptions are inherent in the BDSCV model:

1. The probability of discovering a pool is proportional to its size with an exponential β (i.e., a large pool has a better chance of being discovered).
2. Sampling occurs without replacement (i.e., a pool will not be discovered twice).
3. The pool-size distribution is approximated by a lognormal or nonparametric distribution.
4. The prior distribution of the number-of-pools distribution is approximated by a Poisson distribution or is assigned by geologists.

The first two assumptions are the same as NDSCV, and the first three assumptions are the same as LDSCV. The posterior number-of-pools probability distribution can be any type of distribution. BDSCV provides a probability statement about the N value and also provides a probability measure for each individual pool size (Lee and Wang, 1983b).

Now we use the lognormal hypothetical population and two discovery sequences to demonstrate the advantages of BDSCV. NDSCV was used to make the point estimate about the N value, the nonparametric pool-size distribution, and the exploration efficiency, β. These estimates were entered into BDSCV for estimating the number-of-pools distribution. The Poisson distribution was used as the prior distributions for these examples. BDSCV estimates a posterior distribution based on the input parameters and the discovery sequence for each case. The statistical treatment of the BDSCV model is explained in Appendix A.

The Keg River Shelf

The Rainbow reef play includes all oil trapped in the Keg River pinnacle reefs and the bank-margin reef buildups that accumulated in small deep basins of the Western Canada Sedimentary Basin (Reinson et al., 1993). One hundred sixty-one oil pools have been discovered (Fig. 3.7). The total number of oil pools in this play estimated by NDSCV is 320. After 83 iterations, Figure 3.8 shows the posterior number-of-pools distribution. The expected value of the distribution is 330, and the following probability statement can be made:

$$P(304 \leq N \leq 354) = 0.9$$

The range derived by BDSCV includes the point estimate obtained by NDSCV, but BDSCV presents a range of N values, which provides more information for petroleum resource assessments.

Remarks

BDSCV has been applied to more than 150 oil and gas plays (Table 3.2). In all cases, the posterior distributions cover the point estimates derived by LDSCV or NDSCV, except for a few cases in which LDSCV and NDSCV do not yield a definite answer.

The statistical method of BDSCV is an extension of PETRIMES methodology within the superpopulation framework. Use of BDSCV is made by entering the output from LDSCV or NDSCV to estimate the posterior distribution. On the other hand, geologists can construct a number-of-pools distribution based on the information about the number of prospects and the exploration risk, as shown by Lee and Wang (1983a). The prior number-of-pools distribution (either the Poisson distribution or one constructed by geologists) can be evaluated based on the discovery sequence, and then the posterior number-of-pools distribution can be determined. The assumption of the Poisson distribution has not yet been verified, but BDSCV can still assist petroleum resource assessors by using a distribution for the N value instead of a single

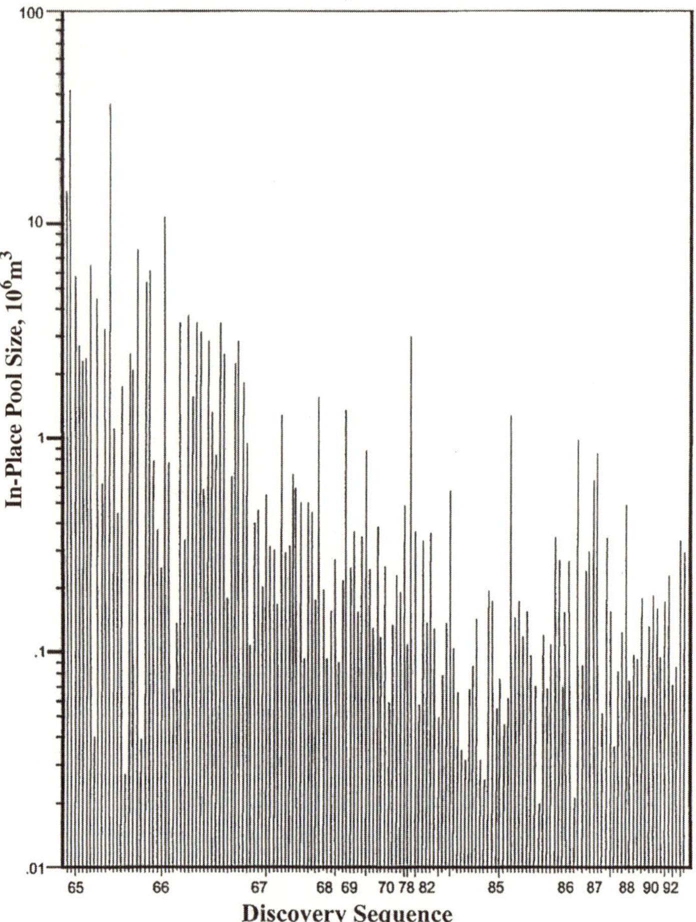

Figure 3.7. Discovery sequence of Keg River–Rainbow reef play, Western Canada Sedimentary Basin. The horizontal axis (no scale) indicates discovery sequence as a function of time (year, month, and day).

point estimate. BDSCV can provide a probability statement about the N value as well as a probability measure of individual pool size. The BDSCV computer program has been implemented in PETRIMES/W (Lee et al., 1999).

Multivariate Discovery Process Model

The order of discovery depends on a combination of attributes, such as deposit volume, water depth, and distance from pipeline. The effect of

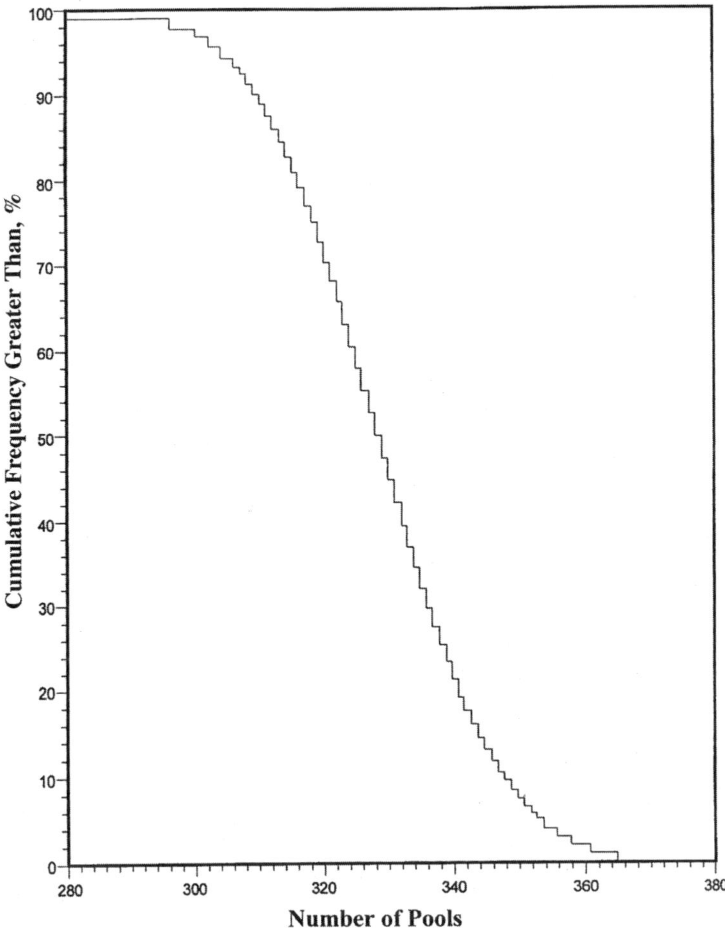

Figure 3.8. Number-of-pools distribution of Keg River–Rainbow reef play, Western Canada Sedimentary Basin. The expected value is 330.

these various factors should be incorporated into the successive sampling discovery model. The magnitude of a pool will therefore depend on multivariate pool attributes: $W_j = W(Y_{j1}, Y_{j2}, \ldots, Y_{jk})$, where the attributes $(Y_{j1}, Y_{j2}, \ldots, Y_{jk})$ associated with a deposit are observable after the deposit is discovered. Specifically, we can write this as follows:

$$W_j = Y_1^{\beta_1} \times Y_2^{\beta_2} \times Y_3^{\beta_3} \ldots \times Y_k^{\beta_j} \quad (3.7)$$

where the β_j's are the parameters to be specified or estimated from the data. For example, Y_1 is the volume, Y_2 is the area, and Y_3 is the porosity

Table 3.2. Examples of Various Plays Characterized by the Parameters μ, σ^2, and N

Population parameters			*Plays*
μ	σ^2	N	
Large	Large	Large	Cretaceous plays, West Siberia Basin Devonian plays, Volga–Ural Basin
Large	Small	Large	Middle Jurassic play, Viking Graben, North Sea
Small	Large	Small	Beaverhill Lake and Rimbey–Meadowbrook reef plays, Western Canada Sedimentary Basin
Small	Small	Large	Mannville plays, Keg River plays, Western Canada Sedimentary Basin
Small	Small	Small	Jurassic and Triassic plays, Western Canada Sedimentary Basin
Small	Large	Large	Devonian clastic play, Western Canada Sedimentary Basin

of a pool. If the porosity has no impact on the order of discovery, β_3 would be close to zero. In principle, many different functional forms can be included here for the magnitude. Time-dependent factors, such as the price of oil, land availability, company, and level of geological knowledge about the play can also be included.

Under the successive sampling discovery model, the play under study is considered to be a statistical population consisting of N oil and gas pools with magnitudes $W_1, W_2, \ldots, W_N$. First, it assumes that the W_1's are generated independently and identically according to a superpopulation distribution F that is assumed to be lognormally distributed in this case. Second, given the magnitudes, W_i, $i = 1, 2, \ldots, N$, the model assumes that the probability of discovering pools with magnitudes $W_1, \ldots, W_n$ first and in that order is $\prod_{j=1}^{n} \frac{W_j}{W_j + \cdots + W_N}$. The magnitude is usually taken to be the area or the volume of a pool.

Let $y_1, \ldots, y_N$ with $y_i = (y_{i1}, \ldots, y_{ik})^T$ be the values associated with the N pools of a finite population that are available for discovery. We assume that the y_i's are generated independently and identically from a multivariate lognormal distribution F_θ with $\theta = (\mu, \Sigma)$. Let F_θ denote the cumulative function and f_θ denote the density function. The aim is to estimate the unknown parameter θ.

Let x_j be the value associated with the jth discovery, and the observed ordered sample be denoted by $(x_1, \ldots, x_n)$. Then the probability of observing the ordered sample $x_1, \ldots, x_n$ under the successive sampling discovery model will be

$$\mathrm{P}\left[(x_1,\ldots,x_n)\middle|(y_1,\ldots,y_N)\right]=\prod_{j=1}^{n}\frac{w(x_j)}{b_j+w(y_{n+1})+\cdots+w(y_N)} \tag{3.8}$$

where $b_j = w(x_j) + \cdots + w(x_n)$. In other words, the sample is obtained by selecting successively, without replacement and with a probability proportional to $w(y)$, from the finite population of N pools.

To obtain the unconditional distribution of the random variable $(X_1, \ldots, X_n)$, we have to sum Equation 3.8 over all possible $\{y_1, \ldots, y_N\}$ values, multiply by the joint density of random variable $Y_1, \ldots, Y_N$, and integrate over the unobserved values $(Y_{n+1}, \ldots, Y_N)$. This gives the joint density of $X_j = x_i,\ \ i = 1, \ldots, n$, as

$$\frac{N!}{(N-n)!}\prod_{j=1}^{n} f_\theta(x_j)\,\mathrm{E}_\theta\left[\prod_{j=1}^{n}\frac{w(x_j)}{b_j+w(Y_{n+1})+\cdots+w(Y_N)}\right] \tag{3.9}$$

Equation 3.9 is the multivariate case of Equation 3.5 and was implemented in PETRIMES/W. The following subsections contain two examples that demonstrate the applications of Equation 3.9.

Bivariate Lognormal Distribution for Oil and Gas Pools

So far, there are three ways to evaluate a trap that contains both oil and gas:

1. Assess the oil and gas separately.
2. Convert the gas into oil-equivalent volume and add it to the oil volume of the same pool.
3. Compute the trap volume for oil and gas together.

The first method presents two assessments, one for oil and one for gas. The second method reports an assessment for oil equivalent. The third method predicts the trap volume.

Equation 3.9 can also be used to estimate the oil and gas joint distribution (Lee, 1998). The following is such an example. The discovery sequence of the Leduc isolated reef play (Fig. 3.9) from the Western

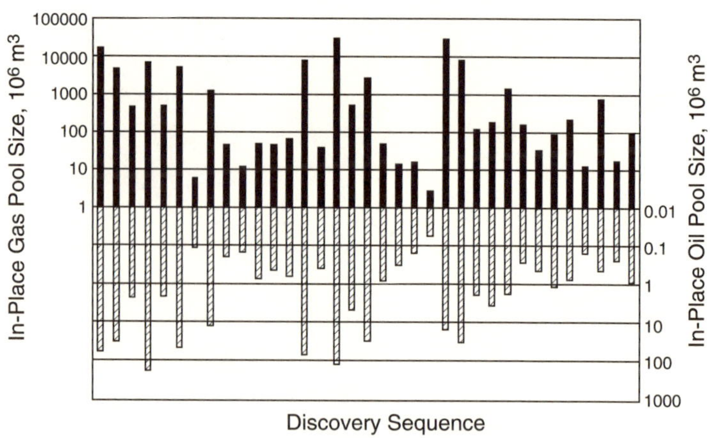

Figure 3.9. Discovery sequence for both oil and gas deposits of the Leduc isolated reef play.

Canada Sedimentary Basin consists of oil and gas within a single reservoir. This example is ideal for demonstrating the application of Equation 3.9 for estimating the bivariate lognormal distribution for both oil and gas deposits.

The estimated mean matrix and covariance matrix for the natural logarithms of gas and oil volumes are $\hat{\boldsymbol{\mu}} = (4.695, -2.560)$ and $\hat{\boldsymbol{\Sigma}} = \begin{pmatrix} 7.155 & 5.973 \\ 5.973 & 5.752 \end{pmatrix}$ respectively. The bivariate lognormal distribution for both gas and oil is

$$f(\text{gas, oil}) = \frac{1}{\text{gas} \times \text{oil} \times 2\pi \times 2.675 \times 2.398 \times \sqrt{1-0.867}} \exp\left\{-\frac{1}{2(1-0.867)} \times \left[\left(\frac{\text{gas}-4.695}{2.675}\right)^2 - 2 \times 0.931 \left(\frac{\text{gas}-4.695}{2.675}\right)\left(\frac{\text{oil}-(-2.560)}{2.398}\right) + \left(\frac{\text{oil}-(-2.560)}{2.398}\right)^2\right]\right\}$$

The oil and gas volumes of each pool have a positive correlation coefficient of 0.867. This can be visualized from Figure 3.9, which shows that a large volume of oil is associated with a large gas volume. The bivariate density function, $f(\text{gas, oil})$, can be used to estimate both gas and oil pool volumes using conditional probability.

Estimating the Covariance Matrix

The reservoir data of the Leduc isolated reef play of the Western Canada Sedimentary Basin was used to demonstrate the impact of biased samples on the estimation of a correlation matrix. The covariance matrix of the random variables—pool area, net pay, porosity, and water saturation—were determined by

1. estimating the values of β and N by LDSCV, NDSCV, or other methods
2. computing the values of β_1, β_2, β_3, and β_4 using multiple regression analysis: log pool size = β_1 log pool area + β_2 log net pay + β_3 log porosity + β_4 log water saturation
3. entering all estimated β_i's, $\hat{N}$, and $\hat{\beta}$ into Equation 3.9 and estimating $\theta = (\boldsymbol{\mu}, \boldsymbol{\Sigma})$

Following are the population covariance matrix estimated by sampling successively from a finite population and the covariance matrix computed by the random sampling assumption (in parentheses). Only the lower diagonal half is shown.

Deposit area	3.061 (2.142)			
Net pay	0.811 (0.472)	0.885 (0.760)		
Porosity	−0.012 (−0.031)	0.068 (0.061)	0.153 (0.152)	
Water saturation	0.018 (0.007)	−0.031 (−0.027)	0.024 (0.023)	0.010 (0.010)

The variances of the random variables, pool area and net pay, are enhanced, as well as the covariance between the pool area and the net pay, which is enhanced from 0.472 to 0.811 if the sampling bias is handled by using the model that samples successively.

Remarks

This section demonstrates that a multivariate discovery process model can be used to estimate the population mean and covariance matrix. Furthermore, the bivariate lognormal pool-size distribution can also be estimated for a play that contains both oil and gas.

Pool-Size-by-Rank by Order Statistics

In resource evaluation, the most useful type of information is the estimation of pool-size-by-rank (the rth largest pools in order statistics), in other words, pool size ranging from largest to smallest. The minimum data required to conduct this operation include (1) a pool-size distribution and (2) the number of pools, N, in the play, or their distribution. The superpopulation concept is assumed for this estimation. Furthermore, the pool-size distribution and the number-of-pools distribution can vary independently, and can be any type of probability distribution (Fig. 3.10).

If $N=1$ (i.e., a single pool play), then the distribution of the largest and smallest pools is precisely given by the pool-size distribution. More generally, if $X_1, X_2, \ldots, X_N$ are pool sizes generated independently from an identical pool-size distribution denoted by F_θ, where $\theta=(\mu,\sigma^2)$, then the greater-than distribution of the largest pool among N pools (Lee and Wang, 1983b) is as follows:

$$L_{N,1}(x)=1-\left[1-F_\theta(x)\right]^N \quad \text{for } X>0 \tag{3.10}$$

The greater-than distribution of the rth largest pool is given by

$$L_{N,r}(x)=\sum_{k=r}^{N}\binom{N}{K}F_\theta(x)\left[1-F_\theta(x)\right]^{N-K} \quad \text{for } X>0,\ r=1,2,\ldots,N \tag{3.11}$$

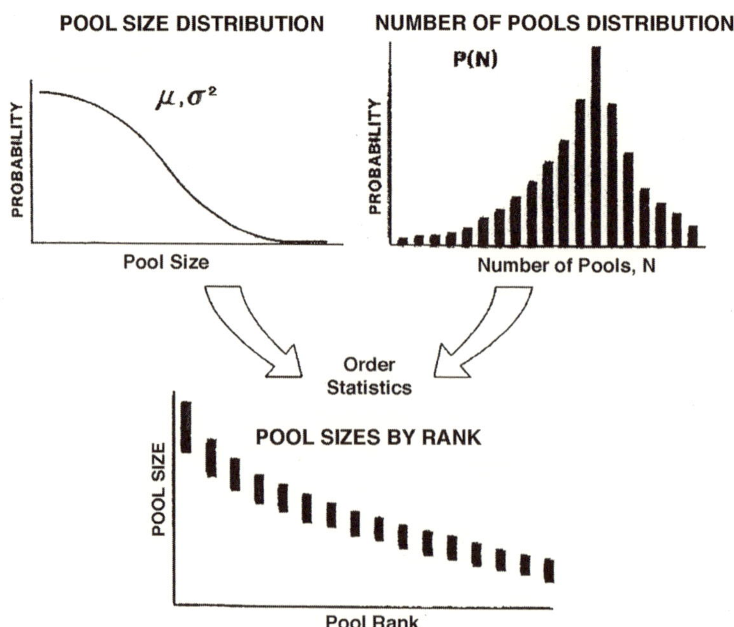

Figure 3.10. Diagram showing the concept of pool-size-by-rank by order statistics.

Equations 3.10 and 3.11 are the distributions of the largest and the rth largest order statistics for a random sample of size N from a superpopulation (Bickel and Doksum, 1977). In petroleum resource evaluation, the density of the rth largest pool can also be derived (Lee and Wang, 1983b) as follows:

$$l_r = \sum_{n=r}^{\infty}\sum_{k=r}^{n}\binom{n}{k}F(x)^k\left[1-F(x)\right]^{n-r} f(x)\mathrm{P}(N=n)/\mathrm{P}(N\geq r) \quad (3.12)$$

for $x > 0$ and $r = 1, 2,\ldots,$ where $\mathrm{P}(N = n)$ is the number-of-pools distribution when $N = n$, and $\mathrm{P}(N \geq r)$ is the number-of-pools distribution when $N \geq r$, for $r = 1, 2,\ldots.$ From Equation 3.12 we see the following:

1. For a fixed set of parameters μ, σ^2, the probability of depositing a largest pool size of at least x increases to 1 as N increases.
2. For a fixed N, and also a given pool size x, the probability of the largest pool being at least x will increase as μ and/or σ^2 increases.

The geological interpretations of these two statements are:

1. If all pools in a play were deposited as a result of the same geological processes (i.e., they are part of the same population), then as the number of pools deposited increases, the more likely it is that one of them will be relatively large.
2. The magnitude of the largest pool tends to change with respect to other pools for different values of μ and σ^2 (i.e., with respect to different geological models). See Appendix C for the statistical treatment.

Interpretations

For the purpose of illustration of pool-size-by-rank, let us reexamine the Beaverhill Lake play. Here, as shown in Figure 3.11, the Swan Hills A pool size (221×10^6 m^3) is located at the upper 1st percentile on the superpopulation pool-size distribution. The interpretation is that the frequency of occurrence of a pool as large or larger than the Swan Hills A pool within the superpopulation is about 1%.

On the other hand, the probability that the largest pool in the Beaverhill Lake play is as large as the Swan Hills A is not 1% but much

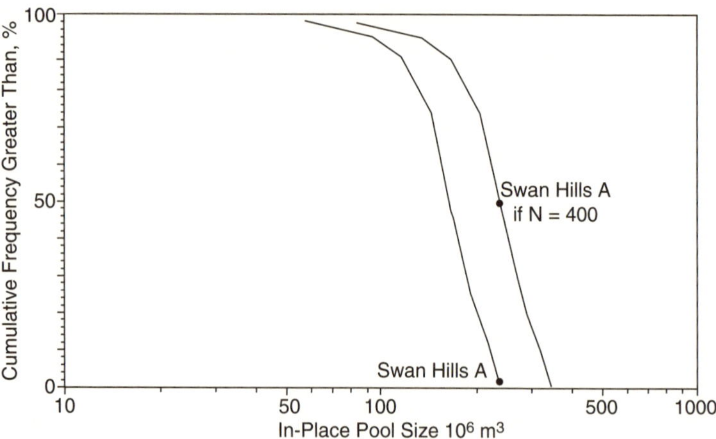

Figure 3.11. Largest pool-size distribution of the Beaverhill Lake play. Note that the largest discovered pool size in Swan Hills A is located at the 50th upper percentile.

higher (unless there is only one pool). In the case of more than one pool, the probability can be obtained from the distribution of the largest pool among N pools. For example, the probability of having the largest pool size as large as Swan Hills A is 0.5 where $N=400$, as shown in Figure 3.11, together with the superpopulation pool-size distribution. In geological terms, given $N=400$, for example, then 400 pools have been deposited with sizes generated from the superpopulation pool-size distribution, and the chance of having the largest of the 400 pools as large as Swan Hills A is 50%. That is to say, if similar geological conditions existed and 400 pools were deposited at one time, then roughly 50% of the time the largest pool would have a size at least as large as that of Swan Hills A. This is a frequentist interpretation of probability that uses the superpopulation concept of pool-size distribution.

The difference in size between two adjacent pools can be examined as a function of σ^2, if N and μ remain unchanged. In Figure 3.12A, the medians of individual pool-size distributions, where $\mu=0.25$, $\sigma^2=6$, and $N=60$, are displayed by dots; the medians of individual pool-size distributions, where $\sigma^2=0.5$ and μ and N remain the same, are displayed by open circles. This figure indicates that pool size decreases more rapidly when σ^2 is relatively large than when σ^2 is relatively small. For any skewed pool-size distribution, such as a lognormal one, given the constant values of μ and N, the larger the value of σ^2, the bigger a

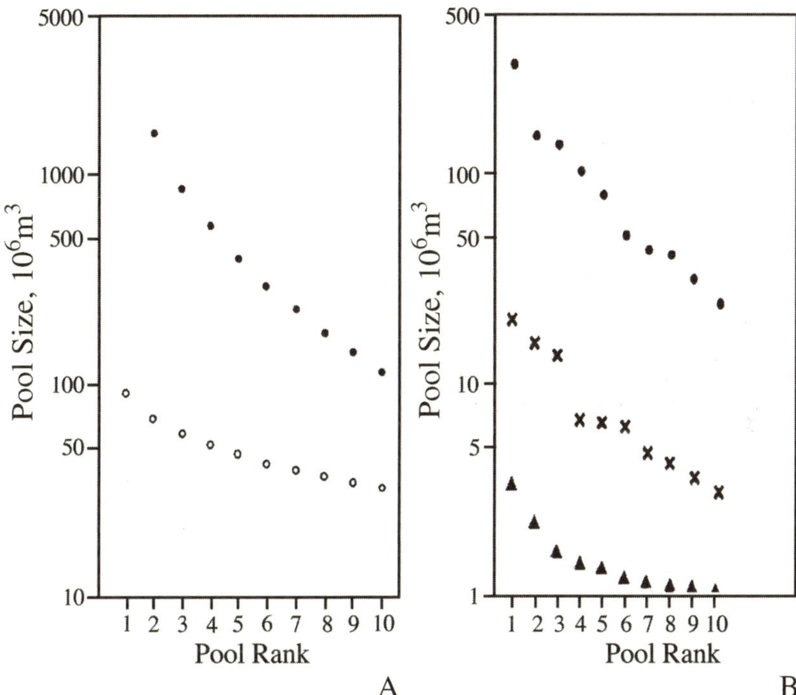

Figure 3.12. (A, B) Median pool-size-by-rank. (A) With population parameters of $\mu = 0.25$ and $N = 60$ for $\sigma^2 = 6.0$ (dots) and $\sigma^2 = 0.5$ (open dots), and (B) for the Beaverhill Lake play (dots) with $\hat{\mu} = 0.25$, $\hat{\sigma}^2 = 6.6$, and $\hat{N} = 60$; the Bashaw play (crosses) with $\hat{\mu} = -0.91$, $\hat{\sigma}^2 = 3.0$, and $\hat{N} = 80$; and the Zama play (triangles) with $\hat{\mu} = -1.5$, $\hat{\sigma}^2 = 1.0$, and $\hat{N} = 160$.

single pool tends to be. Hence the magnitude of the first few large pools among the N pools tends to be greater.

Plays from the Western Canada Sedimentary Basin—the Beaverhill Lake (estimation of this play was based on data that contain commercial pools only), Bashaw, and Zama plays—reveal an interesting pattern. Values of σ^2 were estimated from pool-size data. Figure 3.12B displays the sizes of the largest 10 pools for three plays, which have σ^2 values of 6.6, 3.0, and 1.0 respectively. Note that the estimated variance value of 6.6 for the Beaverhill Lake play was obtained from the data set consisting of commercial pools only. These 10 pools include discovered and undiscovered pools of the plays. The sizes in the Beaverhill Lake play (indicated by dots) decrease more rapidly than those of the Bashaw reef play (indicated by crosses) and those of the Zama play (indicated by triangles). The reason for this change is that the pool-size distribution for the Beaverhill Lake play has the largest variance of all. The reserves from the first 10 pools amount to 91%, 68%, and 46% of their total resources respectively. This phenomenon demonstrates that the magnitude of σ^2 allocates the resources to individual pools.

Distribution of pool-size-by-rank should be computed from either the number of pools, N, or the number-of-pools distribution, and the superpopulation pool-size distribution. The previous discussion may be summarized as follows.

1. The size of the largest pool increases as the number of pools, N, increases. The amount of increase depends on the magnitude of μ and σ^2. For example, pool size increases rapidly when σ^2 is large.
2. In resource evaluation, as we will discuss in Chapter 5, μ dominates σ^2 for the mean of a pool size when the constant is not scaled (see Eqs. 5.8 and 5.9). Therefore, parameters μ and N can be thought of as indicators of the richness of the play, whereas σ^2 and N are indicators of the degree of proneness for having outliers.
3. For each hydrocarbon-bearing play, there is a set of μ, σ^2, and N values associated with the geological model that produced the play. Different geological models can have different values for σ^2, μ, and N, and correspondingly distinct pool sizes. Various play examples are presented in Table 3.2.
4. If a play has a pool-size distribution with a large σ^2, then most of the play's resources will be in the first few largest pools. On the other hand, if σ^2 is relatively small, the pool sizes of the play will be almost equal.
5. Factors that distort estimations of pool-size-by-rank include the problem of mixed populations, errors in estimating the number-of-pools and/or pool-size distributions, and errors in measuring pool sizes. The problem of mixed populations is the most severe one, and causes either under- or overestimation of undiscovered pool sizes when prior distributions are specified. Revisions to the play definition might solve the mixed-population problem. Chapter 4 discusses the impact by mixed populations from simulated data sets.
6. With respect to the significance of changes in the values of N, μ, and σ^2, and their impact on the estimation of individual pool sizes, the largest pool size is sensitive to the following factors (in decreasing importance): σ^2, N, and/or μ.

7. The number of pools in a play, which is a finite number, should include all small pools that might not be economically viable at the time of assessment. One might be concerned that if small pools are included in the assessment, the mean (of the untransformed data) of the pool-size distribution will be substantially reduced. Consequently, the mean would not adequately describe the economic resources. But from the viewpoints of exploration and economic analysis, the remaining largest pool sizes are far more significant than the mean value of the pool-size distribution.
8. As previously mentioned, the superpopulation concept is used by PETRIMES. Thus, the predictions made by the system are of cases that would occur most frequently. A singular case, for example, is that of the Cardium marine sandstone play (see Fig. 2.10), in which the largest pool size is about 10 times larger than the size of the second largest pool. In such a situation, sizes in between the two largest pools may be mistakenly predicted. However, if additional information indicates that no other sizes would exist, then the information can be entered into the system as a condition for predicting the individual pool sizes (see the matching process, discussed in the next section).
9. The concept of pool-size-by-rank can also be explained using Monte Carlo simulation. Assume that we have a pool-size distribution and a number-of-pools distribution. A random number is generated, and the number of pools, N_j (say, 100), is obtained from the number-of-pools distribution. A total of N_j (= 100 in this case) pool sizes is randomly drawn from the pool-size distribution (e.g., $x_1, x_2, \ldots, x_{100}$) and is sorted in descending order. These steps are repeated many times. The largest size from each simulation trial is then used to construct the size distribution of the largest pool. The sizes for the second largest, the third largest, and others are similarly obtained. In practice, the statistical approach for the estimation of individual sizes is more effective and can also provide various matching options.

The Matching Process: Operation

We will now attempt to use the matching process to estimate individual pool-size distributions based on point estimates from either LDSCV or NDSCV. This will assist us in determining N values for immature or conceptual plays. However, before we describe the matching process, a number of prerequisites need to be explained.

Distributions of individual pools can be displayed conveniently without much loss of information as a few selected upper percentiles. Take the pool-size distribution, for example, shown in Figure 3.13. The upper percentiles of 95%, 75%, 50%, 25%, and 5% of the distribution represent 5.5×10^6 m^3, 8.6×10^6 m^3, 11.9×10^6 m^3, 16.4×10^6 m^3, and 26.4×10^6 m^3 of oil respectively. For comparison, the variability of this distribution can be measured by its interquartile range, which measures the variability of the middle 50% of the distribution. In this example, the range for 25% to 75% is given as $16.4 - 8.6 = 7.8$. The larger the interquartile range, the more variable the distribution; hence, the degree of uncertainty will be higher.

There are several reasons why the upper percentiles, as measurements of the individual pool-size distribution, are preferred to the mean and the variance:

1. The mean and standard deviation do not relate directly to probabilities.
2. The mean might overpredict individual pool sizes.

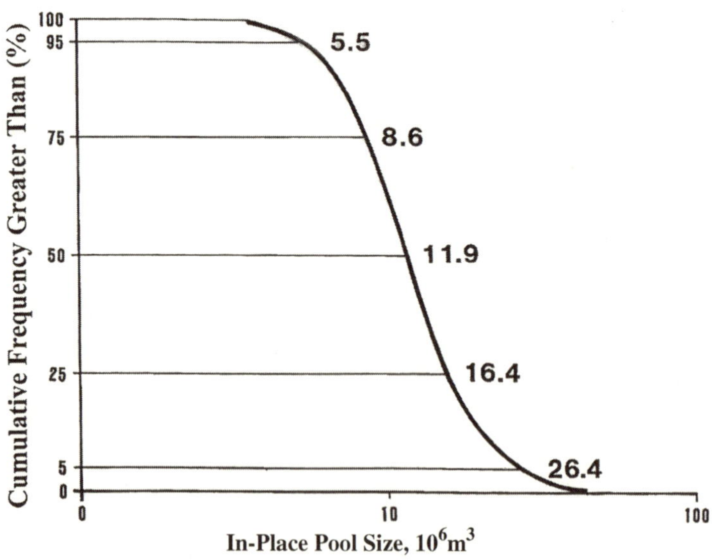

Figure 3.13. Diagram showing meaning of upper percentiles.

3. The standard deviation is typically larger than the mean, which makes it less useful for prediction and comparison.

The interval from the 75th (8.6×10^6 m^3) to the 25th upper percentile (16.4×10^6 m^3) is a 50% prediction interval for the pool that contains the median. That is, the probability that the pool will have a value between 8.6×10^6 m^3 and 16.4×10^6 m^3 is 0.5. Similarly, 5.5×10^6 m^3 to 26.4×10^6 m^3 is a 90% prediction interval for the largest pool. The latter prediction interval has a higher probability of occurrence, but at the expense of having a much wider interval (i.e., more uncertainty).

In the following discussion, we will start with the 75% to 25% prediction interval as a statistical measure of goodness-of-fit, and the median will be used as a point estimator of pool-size-by-rank. The 75% to 25% interval was derived from pilot studies. In cases when the 75% to 25% interval does not match most or all of the discoveries because of the presence of large outliers, the 95% to 5% interval should be used to match current ones.

Matching proceeds as follows: Point estimates derived by LDSCV or NDSCV are used first to predict individual pool sizes because they generally yield reasonable estimates. In cases when the point estimates do not predict all discoveries because variances are underestimated and/or the means are overestimated by LDSCV or NDSCV, the estimated values of μ and σ^2 are chosen from their prediction intervals and reapplied in the matching process to predict all discovered pool sizes. The procedure for finding μ and σ^2 values is as follows:

1. Examine which μ and σ^2 can accommodate most discoveries with the 75th to 25th or 95th to 5th percentile interval.
2. Determine the degree of best prediction by measuring the minimum distance between the discovered pool size and the median or mean of the predicted pool size.
3. Verify statistical predictions by examining their geological implications. After each individual pool-size-by-rank prediction, we can observe whether the implications conform to the geological model.

Examples of the types of question that one should ask after each prediction are: Have we discovered the largest pool yet? What are the sizes of the remaining largest pools? What is the potential of the remaining undiscovered pools? Have we predicted enough small pools for the play? How do recent discoveries, which are not included in the analysis, fit into the prediction picture?

The Beaverhill Lake Play

In the Beaverhill Lake play example, we find that both LDSCV and NDSCV indicate that the value of N may be 400 or more. If $N = 400$ is used, there are three options for choosing a pool-size distribution:

1. The pool-size distribution derived from LDSCV
2. The empirical distribution derived from NDSCV with a mean equal to 2.374×10^6 m^3
3. The lognormal approximation to the empirical distribution with a mean equal to 1.705×10^6 m^3

The matching process is executed based on the empirical distribution. The matching results (Fig. 3.14A) were obtained using the matching process without subjective judgment and are summarized as follows.

1. All pools could be predicted within the 95% to 5% prediction intervals.
2. The discovered pool sizes were matched to the median (upper 50th percentile) of each pool as much as possible.
3. The 13th rank pool could be matched either to the 16th or beyond. We found a number of ways to match the discoveries beyond this rank (Table 3.3).

The matching process can predict no further than this point. Other possible matches were verified using the following procedure:

1. The pool areas corresponding to the remaining largest pool sizes were obtained from the plot of area versus size of the Beaverhill Lake play (Fig. 2.11B).
2. The pool areas obtained in (1) were validated against seismic coverage when seismic grids were small enough to reveal prospects that have the range of pool areas.
3. The predicted pool sizes (Fig. 3.14A) do not show a "smooth" pattern, as do others. This is because the predicted pool-size distributions were derived from the empirical distribution.

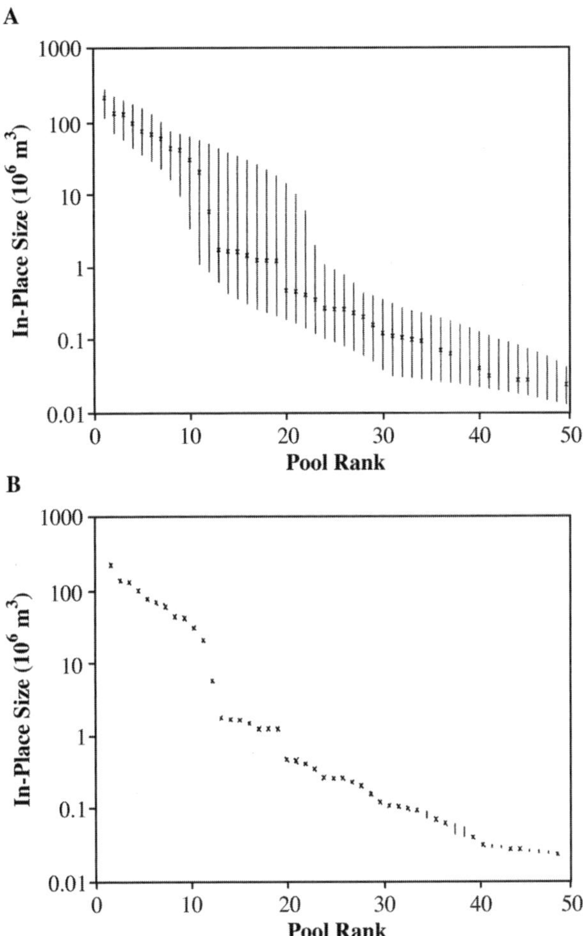

Figure 3.14. (A, B) Median pool-size-by-rank plots derived by nonparametric pool-size distribution and $\hat{N} = 400$. The vertical bars indicate 0.9 probability prediction intervals. The crosses indicate discovered pool sizes. (A) Largest 30 pools displayed with 0.9 probability prediction interval. (B) Largest 30 pools displayed with 0.9 probability prediction interval.

Pool Sizes Conditional on Pool Rank

As indicated in Figure 3.14A, the predicted pool sizes have a wide range of prediction intervals that overlap with the two adjacent pool sizes. This overlapping phenomenon is a result of the uncertainty in the estimations. In this section, we introduce a method to reduce the uncertainty

Table 3.3. The First 20 Pool Sizes Derived by the Matching Process Output

Discovered	*Pool rank*	*Estimated*		
		0.95	*0.50*	*0.05*
211.0	1	116.1	206.1	289.4
134.8	2	70.03	167.4	226.7
130.3	3	56.79	128.4	201.1
98.71	4	44.35	89.23	179.5
76.24	5	36.29	70.50	155.7
69.00	6	29.47	61.92	130.4
60.12	7	22.80	53.33	103.9
44.00	8	16.26	44.74	76.40
41.84	9	9.841	38.49	70.09
31.10	10	3.452	33.42	63.77
21.04	11	1.126	28.36	57.33
5.930	12	0.876	23.30	50.79
—	13	0.628	18.23	44.15
—	14	0.437	13.17	39.24
—	15	0.377	8.11	35.24
1.783	16	0.317	3.04	31.20
1.700	17	0.269	1.16	27.12
1.670	18	0.245	0.96	23.02
1.500	19	0.221	0.75	18.88
1.290	20	0.197	0.55	14.72

by including pool rank in the analysis (Lee and Wang, 1985). The sizes of the undiscovered pools are further constrained by the fact that their size ranges cannot exceed or be less than any discovered (matched) pools that are ranked greater or less than the unmatched pool. The statistical treatment is explained in Appendix D.

Distribution of the Ratio of Two Pools

The distribution of the ratio of two pools can be estimated if the pool-size and number-of-pools distributions are given (see Appendix D). Taking the Beaverhill Lake play as an example, the distribution of the ratio between two pools was estimated and is given in Table 3.4. The ratio leads us to ask the question: Have we discovered the largest pool?

Table 3.4. Examples of Pool Ratios of the Beaverhill Lake Play

Ratios	*Upper percentiles*		
	95	*50*	*5*
1/2	1.106	2.03	16.27
1/3	1.32	3.60	32.33

Play Resource and Potential Distribution

Play Resource Distribution

A geological model can generate a variety of play resource values when using the superpopulation concept. All these resource values constitute the play resource distribution for the superpopulation.

In PETRIMES, the play resource distribution, T, is the sum of all pool sizes of the play. If the pool sizes are approximated using lognormal distributions, then the play resource distributions are the sum of the lognormal distributions. Because this summation does not have an analytical form, the summation is executed using a Monte Carlo simulation procedure. The mean and variance of the play resource distribution are as follows:

$$E[T] = E[X] \times E[N] \tag{3.13}$$

$$\sigma_T = \sigma \times E[N] + (E[X]) \times \sigma_N \tag{3.14}$$

where $E[T]$ is the mean of the play resource distribution, $E[X]$ is the mean of the pool-size distribution, $E[N]$ is the mean of the number-of-pools distribution, σ_T is the variance of the play resource distribution, σ is the variance of the pool-size distribution, and σ_N is the variance of the number-of-pools distribution.

The play resource distribution was derived from the total number of pools, N, and the lognormal or nonparametric pool-size distribution. Figure 3.15B displays the play resource distribution for the Beaverhill Lake play. There is a 90% chance that the play will have a resource ranging from 530×10^6 m^3 to 1670×10^6 m^3, and a 50% chance that the play will have a resource ranging from 789×10^6 m^3 to 1262×10^6 m^3 of oil-in-place. The expected value is 1042×10^6 m^3. The amount of oil that has already been discovered is 947×10^6 m^3.

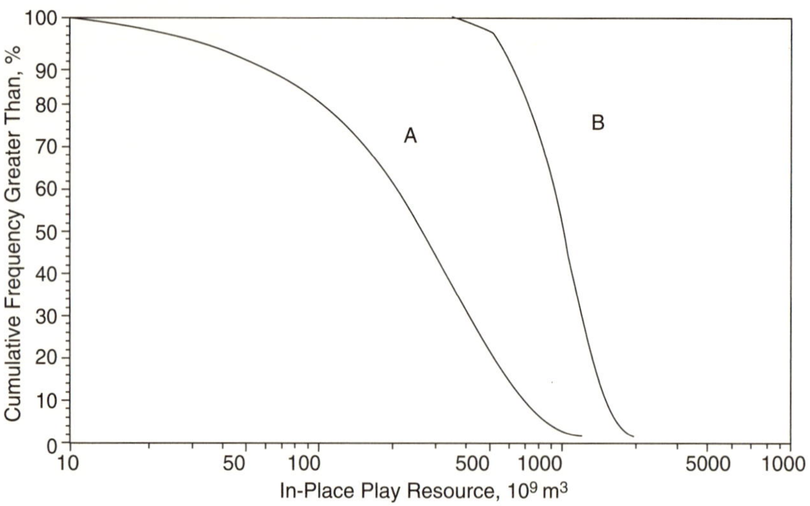

Figure 3.15. (A, B) Potential distribution (A) and play resource distribution (B) for Beaverhill Lake play.

The play resource distribution is the superpopulation distribution and contains the uncertainties explained in the previous chapter. The next two sections describe the remaining play potential.

Play Potential Distribution

Expected Play Potential

Play potential is defined as undiscovered resources that can be estimated from the play resource distribution depending on the matched pool ranks. Summation of the means of all undiscovered pools yields the expected value of the remaining potential distribution and is called the *expected play potential.* The remaining potential is governed by the individual pool sizes and the assigned pool ranks, both of which are determined by the geological play definition used and the quality of the database of the discovered pools. If the discovered sizes are incorrectly estimated, appreciated, or depreciated, or if the rankings are altered, then the expected value of the remaining potential will change. Provided that the geology of a play is well understood and documented, the expected value should provide a reasonable and reliable estimate of the potential of the play.

The play potential distribution (depending on the match, see Fig. 3.14A) shows that there is a 90% chance with the potential ranging from 3.3×10^6 m^3 to 3.5×10^6 m^3, with the mean of the distribution being 3.4×10^6 m^3. Note that the potential distribution is the sum of all undiscovered pools from Figure 3.14A, and therefore is very narrow. A more acceptable range can be derived from the conditional potential, discussed in the next section.

Probable Play Potential Distribution

A conditional play potential can be derived by putting a condition on the amount of discovered resource using the following probability statement:

$$P\left[T = t_1 \middle| T = t_0\right] = \frac{P\left[T = t_1 \cap T = t_0\right]}{P\left[T = t_0\right]} \tag{3.15}$$

where T is the superpopulation resource distribution, t_0 is the amount of discovered resource, and t_1 is the conditional play potential. Equation 3.15 computes the probability of having the conditional play potential, which is referred to as the *probable potential,* including its expected potential. For example, what is the probable (or conditional) potential of the play at a probability of 0.95, given that a total of 949×10^6 m^3 has been discovered? The answer is 26×10^6 m^3.

The Beaverhill Lake Play

After the acceptable match has been estimated, the remaining individual pool sizes and hydrocarbon potential of the play can be estimated by adding conditions to the match. For the Beaverhill Lake play, the remaining pool sizes were estimated by constraining the pool sizes of the 92 discoveries and their ranks. Figure 3.14B and Table 3.3 display the following results:

1. The median of the largest remaining pool sizes can be larger than 3×10^6 m^3 of oil-in-place.
2. The ranges of the prediction intervals derived by the conditional analysis are smaller than those intervals for which conditional analyses were not performed. The overlapping range of two consecutive pool sizes is also much smaller than the case shown in Figure 3.14B.

3. The degree of uncertainty in the prediction intervals is controlled by four factors: (a) the uncertainty inherited from the superpopulation, (b) the ratio of the number of discoveries to total number of pools, (c) the difference in reserves between the two nearest pools, and (d) the individual pool-size distributions computed from specified discovery records, which tend to be less skewed and more concentrated around the medians than those computed without specified conditions.

Estimating pool sizes constrained to a discovery record serves not only to estimate remaining resources, but also to reduce the uncertainty inherited from the superpopulation. The current assessment is different from that conducted in 1988 (Podruski et al., 1988), as follows:

1. The conditional play potential is 26×10^6 m^3 for this assessment, compared with the expected potential of 60×10^6 m^3 in the 1988 assessment.
2. The largest remaining pool size could be as large as 3×10^6 m^3 compared with the 1988 assessment of 7 to 15×10^6 m^3.
3. The total number of pools is 400 in this assessment, compared with 60 in the 1988 assessment.

The reasons for these differences include the following: First, in the 1988 assessment, the Swan Hills A & B pool was considered a single pool with reserves of 303×10^6 m^3. For the current assessment, this pool was divided into two pools (221×10^6 m^3 and 69×10^6 m^3). And second, the minimum pool size adopted by this assessment is about 10 times smaller than that of the 1988 assessment. Therefore, the N value increases accordingly. The current prediction is not distorted because of including noncommercial pools.

4

More about Discovery Process Models

In Chapter 3 we discussed the concepts, functions, and applications of the two discovery process models LDSCV and NDSCV. In this chapter we will use various simulated populations to validate these two models to examine whether their performance meets our expectations. In addition, lognormal assumptions are applied to Weibull and Pareto populations to assess the impact on petroleum evaluation as a result of incorrect specification of probability distributions. A mixed population of two lognormal populations and a mixed population of lognormal, Weibull, and Pareto populations were generated to test the impact of mixed populations on assessment quality. NDSCV was then applied to all these data sets to validate the performance of the models. Finally, justifications for choosing a lognormal distribution in petroleum assessments are discussed in detail.

Validation Study by Simulation

Validation Procedure

Known populations were created as follows: A finite population was generated from a random sample of size 300 ($N = 300$) drawn from the lognormal, Pareto, and Weibull superpopulations. For the lognormal case, a population with $\mu=0$ and $\sigma^2=5$ was assumed. The truncated and shifted Pareto population with shape factor $\theta=0.4$, maximum pool size = 4000, and minimum pool size = 1 was created. The Weibull population with $\lambda=20$, $\theta=1.0$ was generated for the current study. The first mixed population was created by mixing two lognormal populations. Parameters for population I are $\mu=0$, $\sigma^2=3$, and $N_1=150$. For population II, $\mu=3.0$, $\sigma^2=3.2$, and $N_2=150$.

The second mixed population was generated by mixing lognormal ($N_1=100$), Pareto ($N_2=100$), and Weibull ($N_3=100$) populations with a total of 300 pools. In addition, a gamma distribution was also used for reference.

The lognormal distribution is J-shaped if an arithmetic scale is used for the horizontal axis, but it shows an almost symmetrical pattern when a logarithmic scale is applied. The probability density function of a lognormal distribution is defined as

$$f(x)=\frac{1}{\sigma x\sqrt{2\pi}}\exp\left[-\frac{1}{2}\frac{\ln x-\mu)^2}{\sigma^2}\right] \quad (4.1)$$

where x is the pool size, μ is the mean of the logarithmic transformed data, and σ^2 is the variance of the logarithmic transformed data.

The Weibull population displays a J-shaped distribution if the data are plotted on an arithmetic scale, whereas it is almost symmetric but skewed toward the left when plotted on a logarithmic scale. The probability density function of a Weibull distribution is defined as

$$f(x)=\frac{\alpha}{\beta^\alpha}x^{(\alpha-1)}\exp\left[-\left(\frac{x}{\beta}\right)^\alpha\right] \quad (4.2)$$

where x is the pool size, with α (shape factor) > 0, and β (spread factor) > 0.

The histograms of gamma and Pareto distributions display J-shaped distributions on both arithmetic and logarithmic scales. The probability density function of a gamma distribution is defined as

$$f(x)=\frac{x^{\alpha-1}\exp\left(-\frac{x}{\beta}\right)}{\beta^\alpha\Gamma(\alpha)} \quad (4.3)$$

where x is the pool size, with σ (shape factor) >0, β (spread factor) > 0.

The truncated and shifted probability density function of the Pareto distribution is defined as

$$f(x)=\frac{\theta x^{-(\theta-1)}}{a^{-\theta}-b^{-\theta}} \tag{4.4}$$

where x is the pool size, a is the lower limit of the pool size, b is the upper limit of the pool size, and θ is the shape factor. The tested populations are shown in figures 4.1 and 4.2.

Populations were generated for lognormal, Weibull, Pareto, mixtures of two lognormals, and mixtures of lognormal, Weibull, and Pareto populations. The discovery sequences for each of these populations were simulated (using $\beta=0.6$) and are shown at the top of figures 4.3 through 4.7. For each sequence, various numbers of pools are also discovered (given, in this example, values of $n=30$ and $n=50$).

LDSCV and NDSCV were then used to analyze each of these discovery sets to examine whether we can predict the known populations. The following sections discuss the reliability of the assessments derived from both discovery process models based on the following estimated results: N value, β value, pool-size-by-rank, and play resource distributions.

Estimates for the *N* Value

A discovery sequence contains information about the total number of pools in a play as expressed by the LDSCV model (Eq. 3.5) and the NDSCV model (Appendix B). The reliability of estimating N can be validated using the tested populations with the known population mean and variance and the total number of pools. Although the results are based on a single simulation trial, the interpretations can be applied to similar cases.

Lognormal Population

In an ideal situation, the log-likelihood value should show a maximum value from which N could be determined. This relationship may show a negative exponential curve when the ratio value n/N and/or β value is small. In these examples, the log-likelihood values versus N show negative exponential curves, but the curves flatten when $N=300$ for

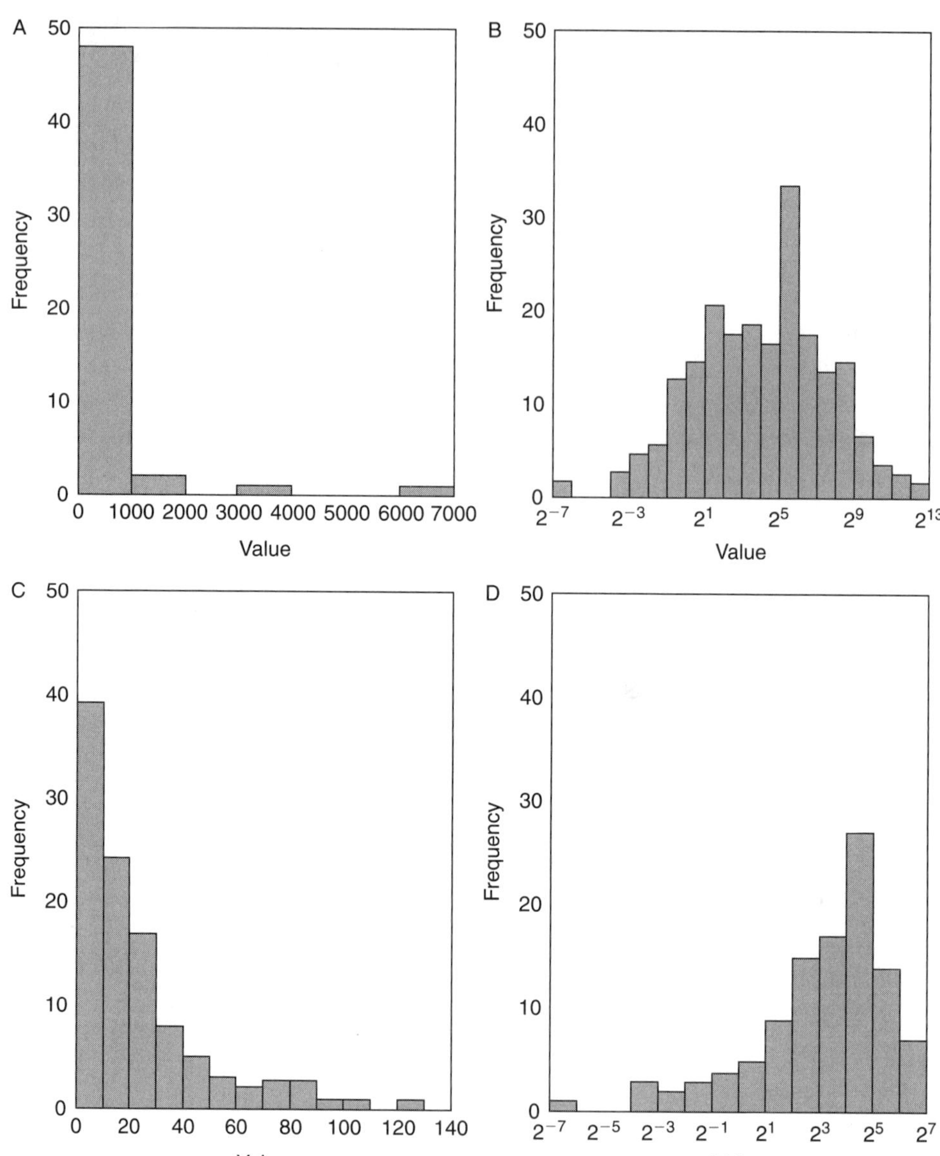

Figure 4.1. (A–D) J-shaped histograms of the lognormal population (A) and Weibull population (C) plotted on a linear scale. Bell-shaped histograms of the lognormal population (B) and Weibull population (D) plotted on a logarithmic scale (base 2).

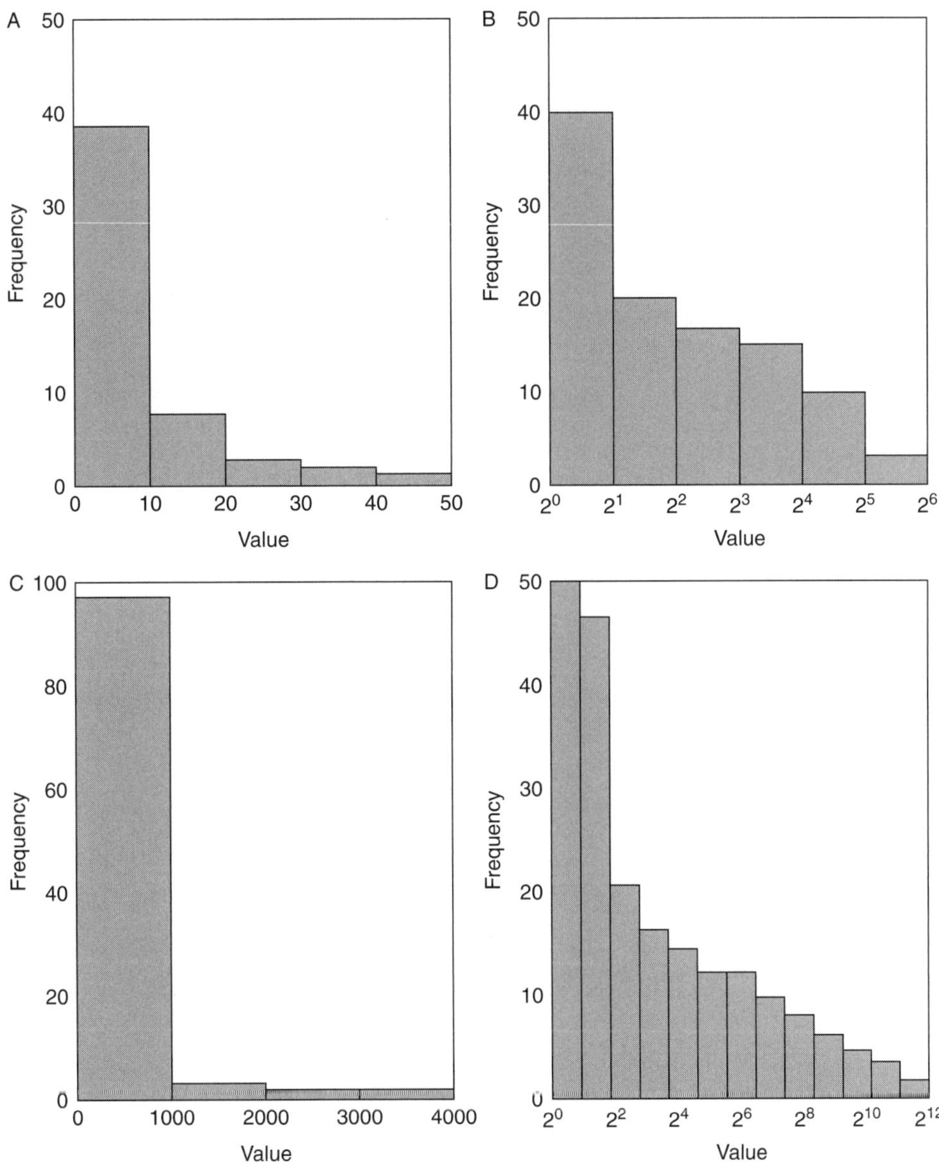

Figure 4.2. (A–D) J-shaped histograms of the gamma population plotted on a linear scale (A) and a logarithmic scale (B). Pareto population plotted on an arithmetic scale (C) and a logarithmic scale (D).

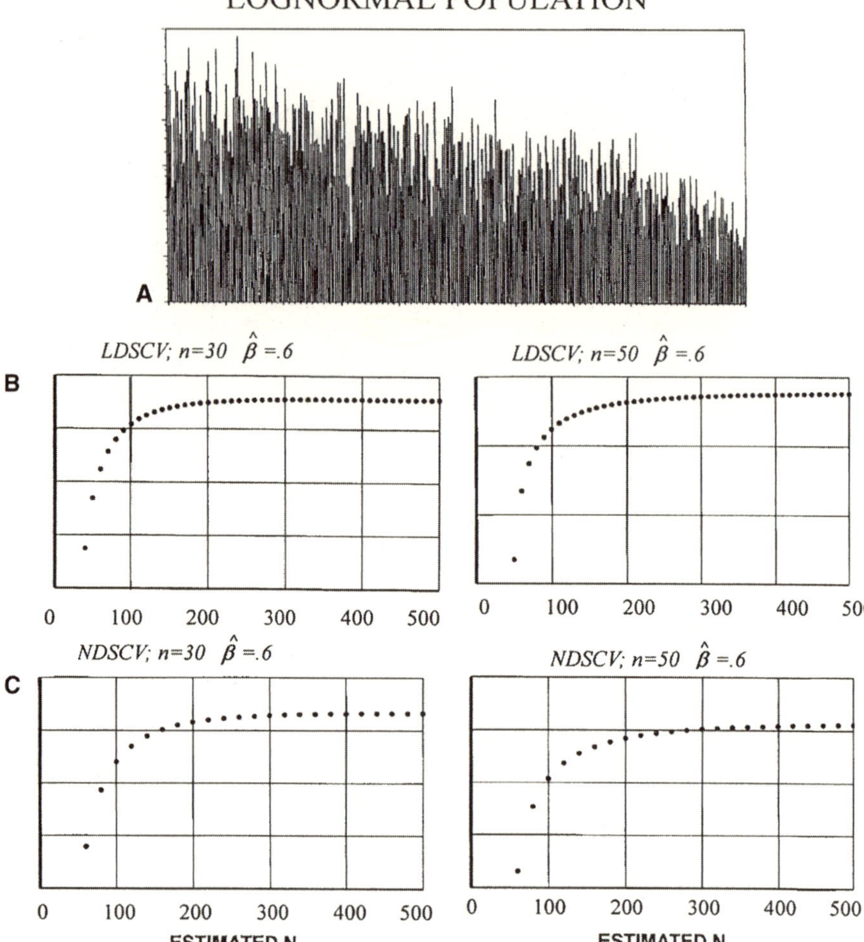

Figure 4.3. (A–C) Simulated lognormal population. Discovery sequence (A) and log-likelihood values versus N value plots derived by LDSCV (B) and NDSCV (C).

both LDSCV and NDSCV. The LDSCV and NDSCV results are given in Figure 4.3 and tables 4.1 through 4.4. More information about the procedure for determining N values can be found in Lee et al. (1999).

Weibull Population

For the Weibull population, LDSCV underestimates the number of pools as 200 when $n = 30$ (Fig. 4.4B, left; Table 4.1) and overestimates the number of pools as 400 when $n = 50$ (Fig. 4.4B, right; Table 4.2). On

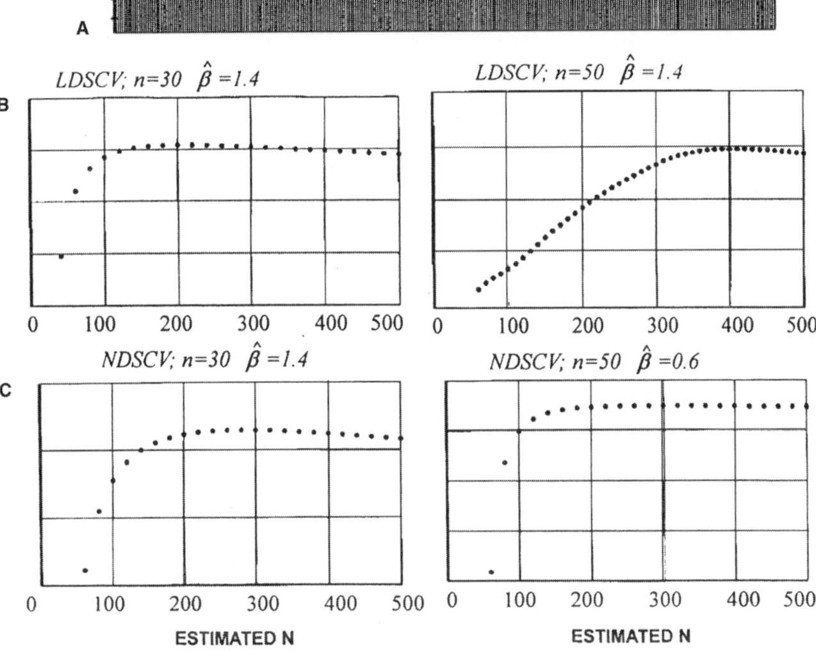

Figure 4.4. (A–C) Simulated Weibull population. Discovery sequence (A) and log-likelihood values versus N value plots derived by LDSCV (B) and NDSCV (C).

the other hand, NDSCV shows that $\hat{N}=280$ when $n=30$, and $\hat{N}=300$ when $n=50$ (Fig. 4.4C; tables 4.3 and 4.4).

Pareto Population

For the Pareto population, LDSCV underestimates the number of pools as 220 when $n=30$ (Fig. 4.5B, left; Table 4.1), but when $n=50$,

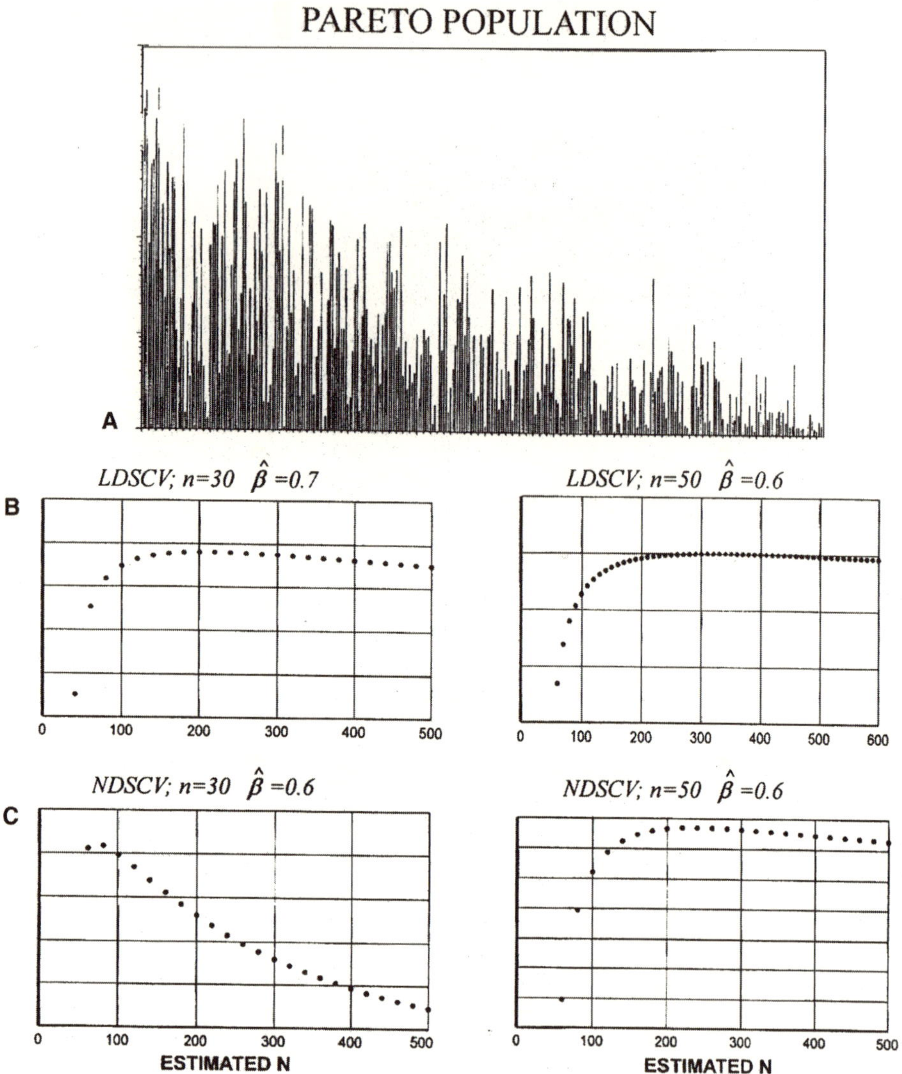

Figure 4.5. (A–C) Simulated Pareto population. Discovery sequence (A) and log-likelihood values versus N value plots derived by LDSCV (B) and NDSCV (C).

$\hat{N} = 300$ (Fig. 4.5B, right; Table 4.2). For the NDSCV case, it underestimates the number of pools as 220 and 260 when $n = 30$ (Fig. 4.5C, left; Table 4.3) and $n = 50$ (Fig. 4.5C, right; Table 4.4) respectively.

Mixed Population of Two Lognormal Populations

The LDSCV log-likelihood values show a maximum value at $\hat{N} = 100$ (Fig. 4.6B, left; Table 4.1) when $n = 30$. When $n = 50$, the log-likelihood

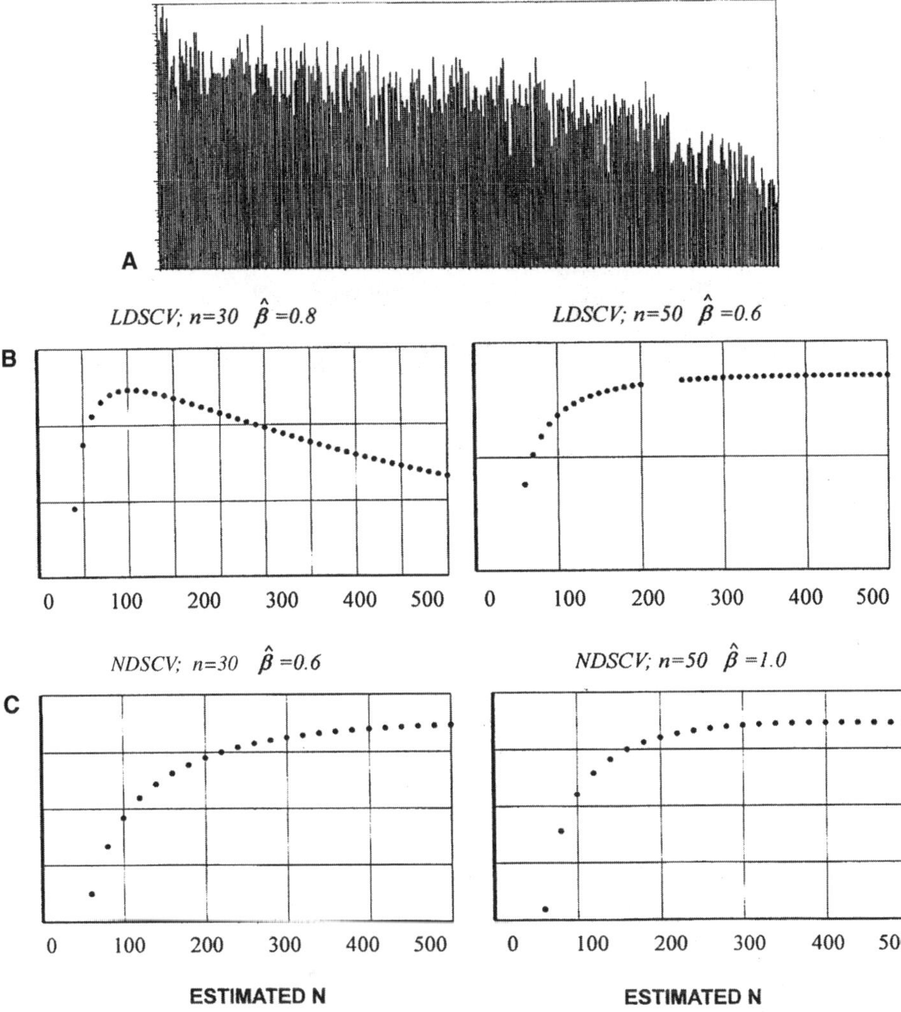

Figure 4.6. (A–C) Simulated mixed population of two lognormal populations. Discovery sequence (A) and log-likelihood values versus N value plots derived by LDSCV (B) and NDSCV (C).

values show a negative exponential relationship and reach a plateau at $\hat{N} = 300$ (Fig. 4.6B, right; Table 4.2). On the other hand, NDSCV yields $\hat{N} = 300$ when $n = 30$ and 50 (Fig. 4.6C, tables 4.3 and 4.4).

Mixed Population of Lognormal, Weibull, and Pareto Populations

The LDSCV log-likelihood values do not show an N value when $n = 30$, and therefore, $\hat{N} = 300$ is assumed (Fig. 4.7, Table 4.1). On the other

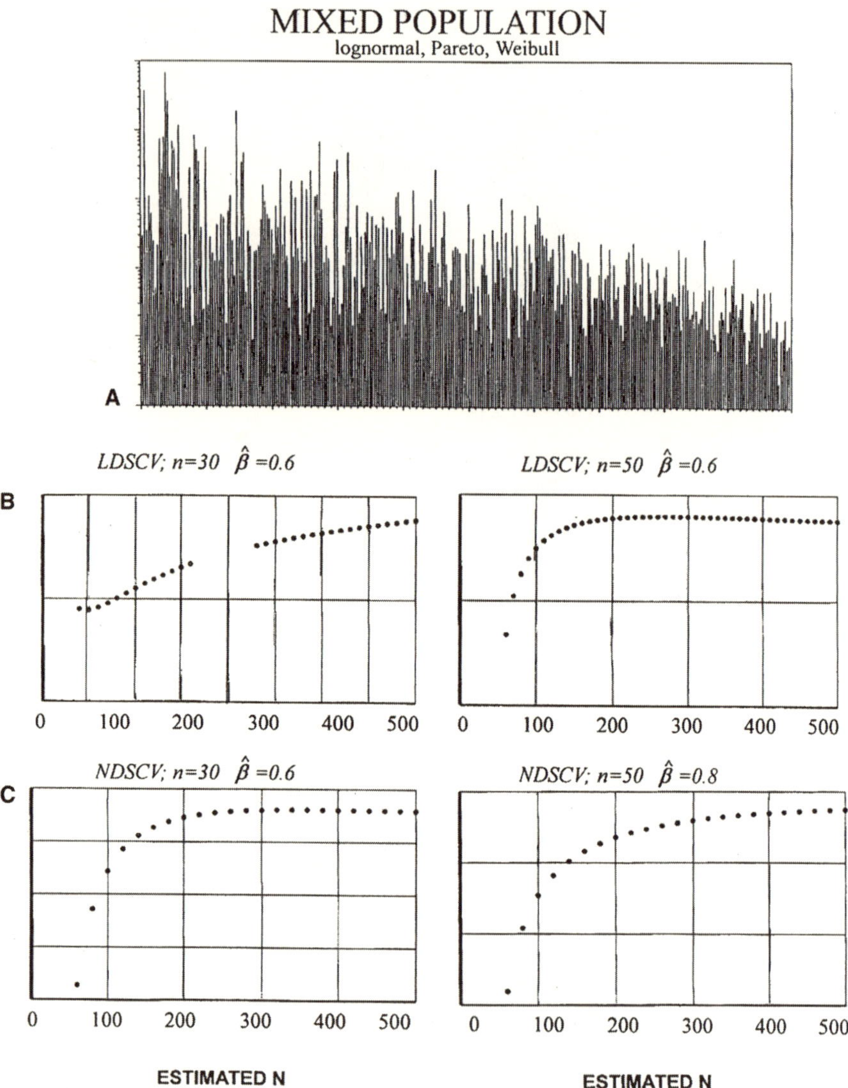

Figure 4.7. (A–C) Simulated mixed population of lognormal, Weibull, and Pareto populations. Discovery sequence (A), and log-likelihood values versus N value plots derived by LDSCV (B) and NDSCV (C).

hand, the LDSCV log-likelihood values show a maximum of 300 when $n = 50$ (Fig. 4.7B, right; Table 4.2). The NDSCV log-likelihood values yield a maximum at 300 when $n = 30$ (Fig. 4.7C, left; Table 4.3); but when $n = 50$, likelihood values show a negative exponential curve (Fig. 4.7C, right; Table 4.4) and reach a plateau at $\hat{N} = 300$.

Table 4.1. Summary of the Estimates for Various Populations When $n = 30$ (Lognormal Assumption Is Used)

Types of populations	*Total resources*	*N*	$\hat{N}$	$\hat{\beta}$	*Upper percentiles*				
					95	*75*	*50*	*25*	*5*
Lognormal	50,901	300	300	0.6	29,549	36,507	43,390	53,599	77,764
Weibull	6100	300	200	1.4	2231	2478	2682	2887	3246
Pareto	30,375	300	220	0.7	6958	13,000	22,533	45,569	183,400
Mixture of two lognormals	35,526	300	100	0.8	8974	14,884	21,952	34,633	88,884
Mixtures of lognormal, Weibull, and Pareto	32,333	300	300	0.6	19,845	27,342	36,067	51,308	98,837

Table 4.2. Summary of the Estimates for Various Populations When $n = 50$ (Lognormal Assumption Is Used)

Types of populations	*Total resources*	*N*	$\hat{N}$	$\hat{\beta}$	*Upper percentiles*				
					95	*75*	*50*	*25*	*5*
Lognormal	50,901	300	300	0.6	42,921	51,586	59,560	70,796	95,843
Weibull	6100	300	400	1.0	5311	5729	6045	6391	6977
Pareto	30,375	300	300	0.7	14,547	20,714	28,149	41,694	87,147
Mixture of two lognormals	35,526	300	300	0.6	25,279	30,498	35,369	42,220	57,813
Mixtures of lognormal, Weibull, and Pareto	32,333	300	300	0.6	16,114	21,871	28,568	39,927	74,302

From the results of these simulation studies, it can be concluded that when the number of discoveries, n, increases, the impact of the shape of a probability distribution diminishes. One can also conclude that the mixture of two lognormal populations with different means and variances does not significantly distort the estimation of N as sample size increases. In addition, if a mixed population consists of various probability distributions, then NDSCV can be used, whereas LDSCV might provide information about the N value as the sample size, n, increases.

Table 4.3. Summary of the Estimates for Various Populations When $n = 30$ (Lognormal Assumption Is *Not* Used)

Types of populations	*Total resources*	*N*	$\hat{N}$	$\hat{\beta}$	*Upper percentiles*				
					95	*75*	*50*	*25*	*5*
Lognormal	50,901	300	300	0.6	27,944	33,549	38,456	45,181	57,801
Weibull	6100	300	280	1.4	1983	2228	2405	2599	2883
Pareto	30,375	300	220	0.6	9241	14,119	17,995	22,179	29,212
Mixture of two lognormals	35,526	300	300	1.2	12,300	19,295	25,412	32,851	44,841
Mixtures of lognormal, Weibull, and Pareto	32,333	300	300	0.6	20,099	26,642	32,324	39,050	49,842

Table 4.4. Summary of the Estimates for Various Populations When $n = 50$ (Lognormal Assumption Is *Not* Used)

Types of populations	*Total resources*	*N*	$\hat{N}$	$\hat{\beta}$	*Upper percentiles*				
					95	*75*	*50*	*25*	*5*
Lognormal	50,901	300	300	0.6	40,813	49,407	56,926	66,014	80,416
Weibull	6100	300	300	0.6	5563	5939	6221	6493	6913
Pareto	30,375	300	260	0.6	16,629	22,405	26,504	31,287	38,849
Mixture of two lognormals	35,526	300	300	1.0	18,474	25,577	32,599	43,826	56,071
Mixtures of lognormal, Weibull, and Pareto	32,333	300	300	0.8	18,182	24,792	30,623	37,371	48,040

Estimation of Exploration Efficiency

Exploration efficiency measures how fast explorationists can discover the few largest pools in a play. During the past several decades, a number of methods for estimating exploration efficiency have been suggested. The history of estimating exploration efficiency merits a brief review.

Drilling efficiency, C, as defined by Arps and Roberts (1958) is discussed in Chapter 7 (see Eq. 7.4). Arps and Roberts classified the reasons for drilling a prospect into three classes and proposed that if the drilling is conducted randomly on a trend, then $C=1$. If the drilling decision is based on geological and geophysical leads, then $C=2$. The third class lies between one and two. Drew et al. (1980) and Drew (1990) adopted the Arps and Roberts method and used past exploration efficiency for predictions of the immediate future in the study of the Denver Basin and the Permian Basin of West Texas and southeastern New Mexico.

Drew (1990; Drew et al., 1980) used the following approach for selecting a C value for a particular depth interval. The procedure is to carry out a retrospective study. For example, in the 1961 to 1974 forecasts, the total oil and gas combined equals the actual discoveries within the same period, so that the C value obtained equals two. In the Gulf of Mexico offshore study, Drew et al. (1982) used a nonlinear regression method to estimate simultaneously the number of fields and the efficiency of exploration for each size class. The efficiency of exploration ranged from 2.55 to 5.35. This method was used by Arps and Roberts (1958) and by Drew (1990; Drew et al., 1980, 1982) to forecast the future discovery rate based on the C value obtained.

Bloomfield et al. (1979) used the Monte Carlo procedure to estimate discoverability for a Kansas data set and obtained a discoverability coefficient of 0.3. Forman and Hinde (1985) found an empirical straight-line relationship between the logarithmic hydrocarbon volume and the number of fields, N, as

$$\log V = a + bN \tag{4.5}$$

where a is the intercept and b is a negative value for the slope of the fitted line. The ability of the explorationist to discover larger pools first is specified by the slope, b. The greater the degree to which larger pools are discovered first, the steeper the slope.

The purpose of using PETRIMES to estimate the β value is to account for other factors that are not included in the likelihood function of Equation 3.5 and to obtain the mean and variance of the pool-size distribution. Two procedures can be used: (1) with LDSCV, N can be obtained by the maximized β value; and (2) with LDSCV or NDSCV, a specific value can be assigned to β and the log likelihood is computed. By selecting the highest log-likelihood value, the plausible value of β can then be chosen.

For the lognormal case, both LDSCV and NDSCV can predict the β values correctly (Fig. 4.3, tables 4.1 through 4.4). For the Weibull case, LDSCV overestimates the β value (Fig. 4.4, tables 4.1 and 4.2) when $n = 30$ and 50, whereas NDSCV overestimates its value when $n = 30$ (Fig. 4.4, Table 4.3), and yields a correct estimate when $n = 50$ (Fig. 4.4, Table 4.4). For the Pareto example, LDSCV overestimates the β value when $n = 30$ and yields a correct estimate when $n = 50$. On the other hand, NDSCV presents the correct estimates when $n = 30$ and 50. For the mixed lognormal population case, LDSCV overestimates the β value when $n = 30$ and presents a correct estimate when $n = 50$ (Fig. 4.6, tables 4.1 and 4.2), whereas NDSCV overestimates the β values when $n = 30$ and 50 (Fig. 4.6, Tables 4.3 and 4.4). For the mixed population of lognormal, Weibull, and Pareto populations, LDSCV gives the correct estimates when $n = 30$ and 50 (Fig. 4.7, tables 4.1 and 4.2), whereas NDSCV presents a correct estimate when $n = 30$, and overestimates its value when $n = 50$ (Fig. 4.7, tables 4.3 and 4.4).

Pool-Size-by-Rank

The point estimates derived by LDSCV and the empirical distributions derived by NDSCV were used to compute the pool-size-by-rank for all cases. We shall examine the plots for each case. For the lognormal cases, LDSCV (Fig. 4.8A, B) and NDSCV (Fig. 4.8C, D) can predict all pools within the 0.9 probability prediction intervals. For the Weibull case, both LDSCV and NDSCV can predict the largest six pools (Fig. 4.9A, C), but cannot predict the rest of the pools when $n = 30$. When $n = 50$, LDSCV can predict the first 20 pools (Fig. 4.9B), and NDSCV can predict all pools (Fig. 4.9D) when $n = 50$. For the Pareto case, when $n = 30$, both LDSCV and NDSCV can predict the first eight largest pools (Fig. 4.10A) and the first 14 largest pools (Fig. 4.10C) respectively. When $n = 50$, both LDSCV (Fig. 4.10B) and NDSCV (Fig. 4.10D) can predict all pools within the 0.9 probability prediction interval.

For the mixed population cases, LDSCV predicts all pools when $n = 30$ and 50 (Fig. 4.11A, B), but NDSCV can only predict the first 17 pools when $n = 50$ (Fig. 4.11C, D). It is obvious that LDSCV performs better than NDSCV if the mixed population is made up of lognormal distributions. For the mixed population of lognormal, Weibull, and Pareto populations, both LDSCV and NDSCV can predict all pools when $n = 30$ and 50 (Fig. 4.12).

Play Resource Distribution

Play resource distributions for all cases derived by LDSCV and NDSCV were computed (tables 4.1 through 4.4). The 0.9 probability

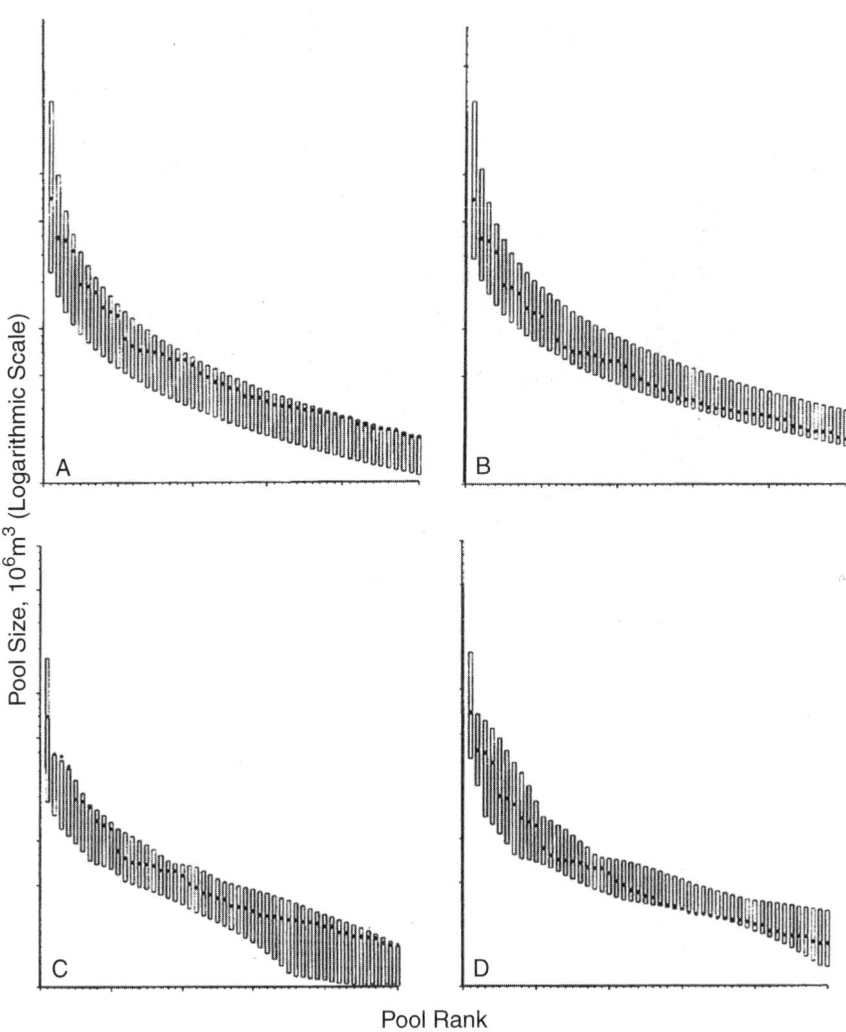

Figure 4.8. (A–D) Pool size-by-rank plots for a lognormal population derived by LDSCV when $n = 30$ (A) and $n = 50$ (B), and plots derived by NDSCV when $n = 30$ (C) and $n = 50$ (D). The prediction interval is the 0.9 probability level.

prediction interval was used to measure the performance ability of LDSCV and NDSCV (i.e., if the interval could include the population value, then the predictions were acceptable). LDSCV predicts all the population values within the interval except the Weibull population when $n = 30$. As n increases to 50, the LDSCV intervals include all population values.

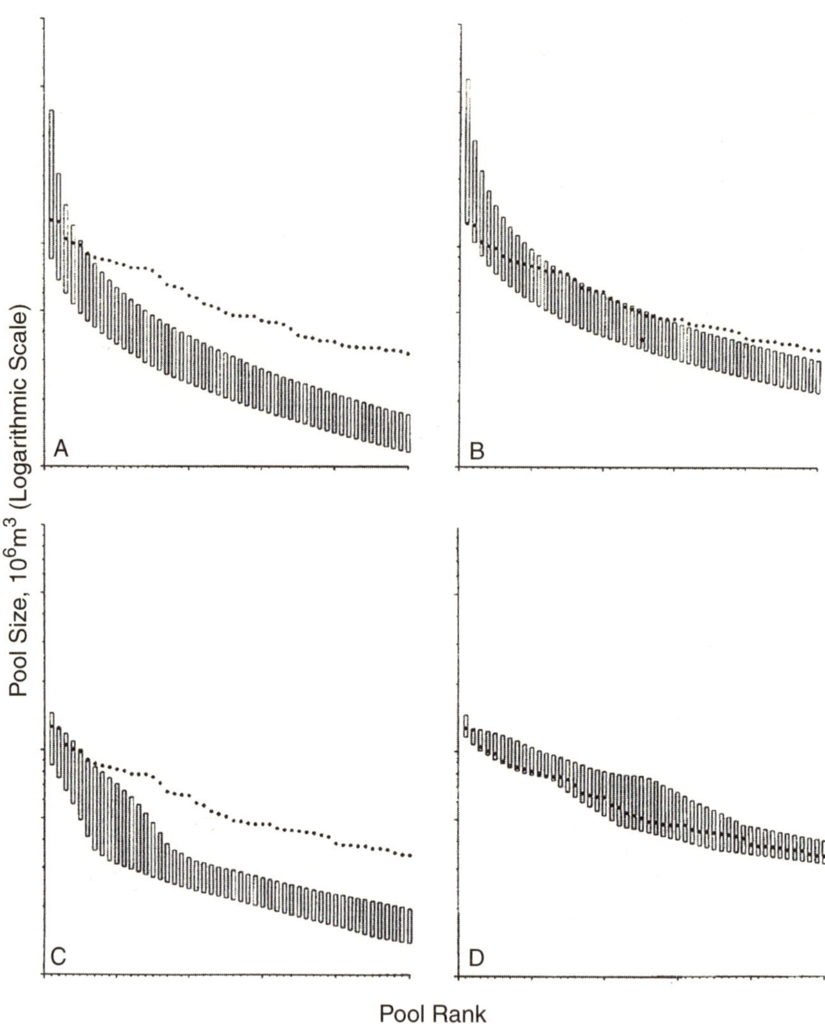

Figure 4.9. (A–D) Pool-size-by-rank plots for a Weibull population derived by LDSCV when $n = 30$ (A) and $n = 50$ (B), and plots derived by NDSCV when $n = 30$ (C) and $n = 50$ (D). Prediction interval is the 0.9 probability level.

On the other hand, the NDSCV prediction intervals include the population values of the lognormal, the mixed population of two lognormal populations, and the mixed population of the lognormal, Weibull, and Pareto populations when $n = 30$, but cannot include the population values of the Weibull and Pareto populations. Again, as n increases to 50, NDSCV can predict all population values within the interval.

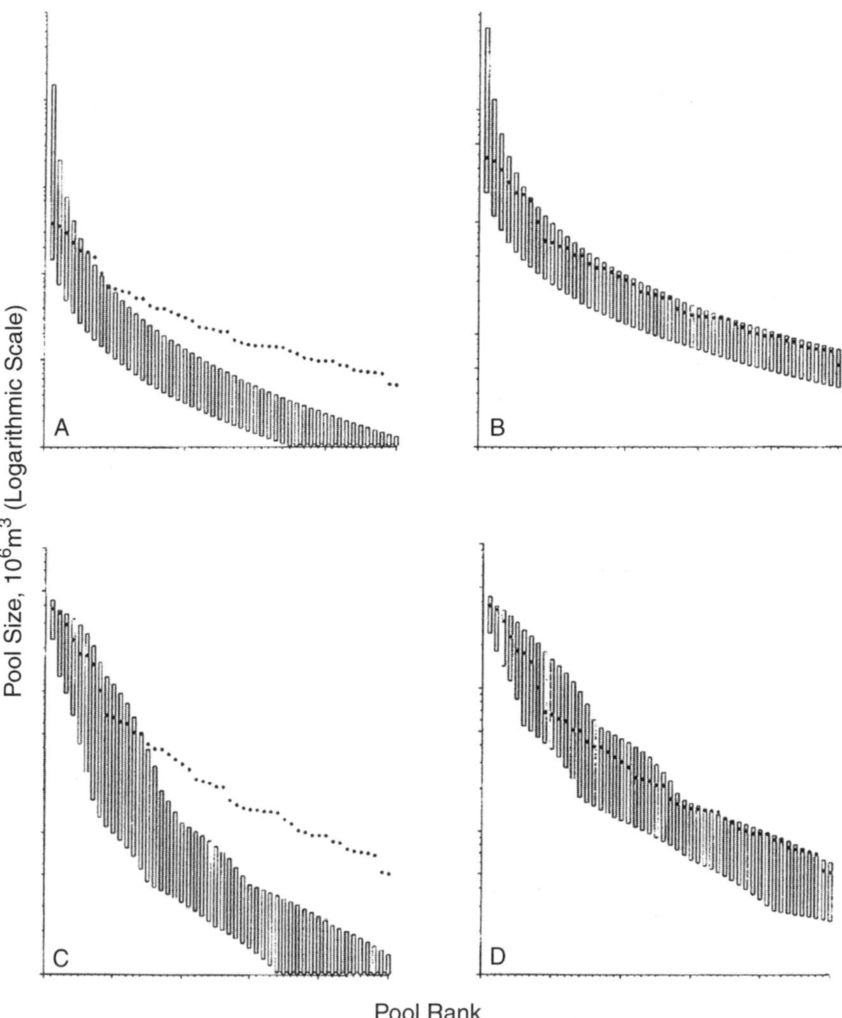

Figure 4.10. (A–D) Pool-size-by-rank plots for a Pareto population derived by LDSCV when $n = 30$ (A) and $n = 50$ (B), and plots derived by NDSCV when $n = 30$ (C) and $n = 50$ (D). Prediction interval is the 0.9 probability level.

Reduction of Uncertainty

With both LDSCV and NDSCV methods, estimation uncertainty decreases when sample size increases, as demonstrated by the following procedure.

1. A random sample of size N (= 300) was drawn from the superpopulation ($\mu = 0.0$ and $\sigma^2 = 5.0$).

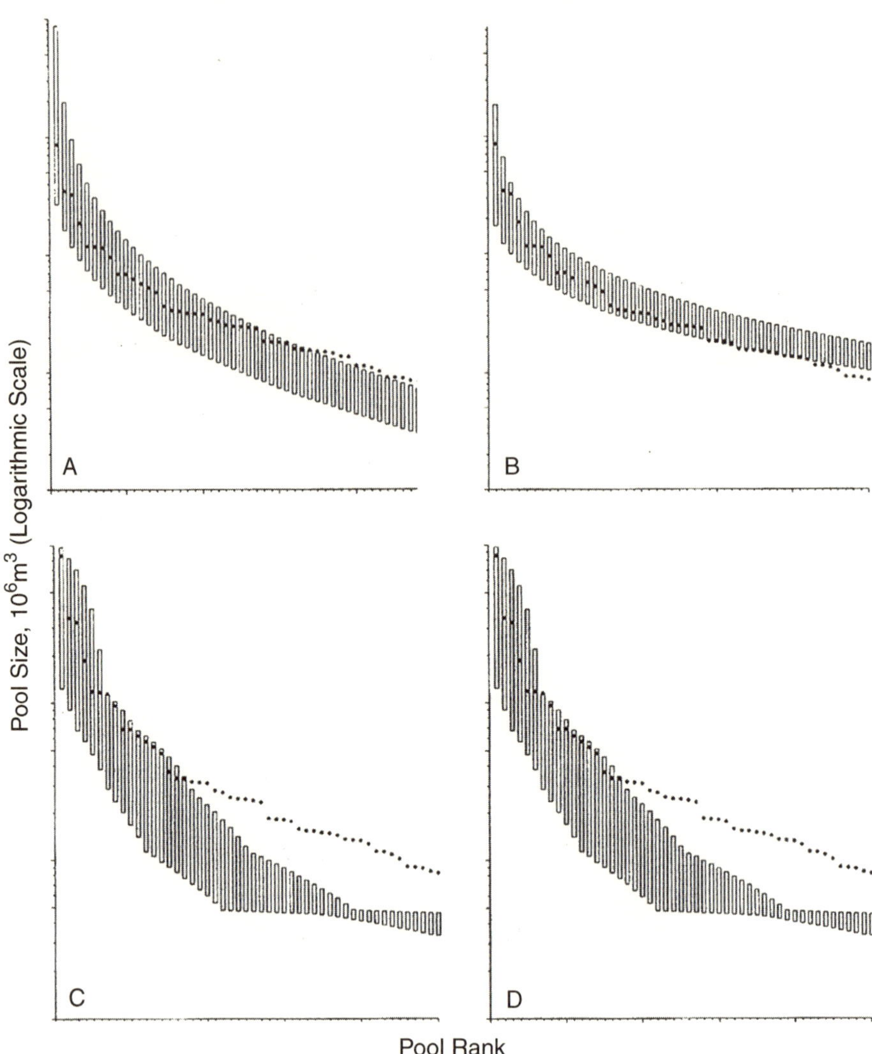

Figure 4.11. (A–D) Pool-size-by-rank plots for mixed population of two lognormal populations derived by LDSCV when $n = 30$ (A) and $n = 50$ (B), and plots derived by NDSCV when $n = 30$ (C) and $n = 50$ (D). Prediction interval is the 0.9 probability level.

2. A discovery process was simulated with a sample of size n (30, 50, 100, and 150) with $\beta = 0.6$.
3. Samples obtained from these simulations were analyzed using LDSCV and NDSCV.
4. Steps 1 through 3 were repeated 1000 times, so 1000 pairs of estimated μ and σ^2 were obtained.

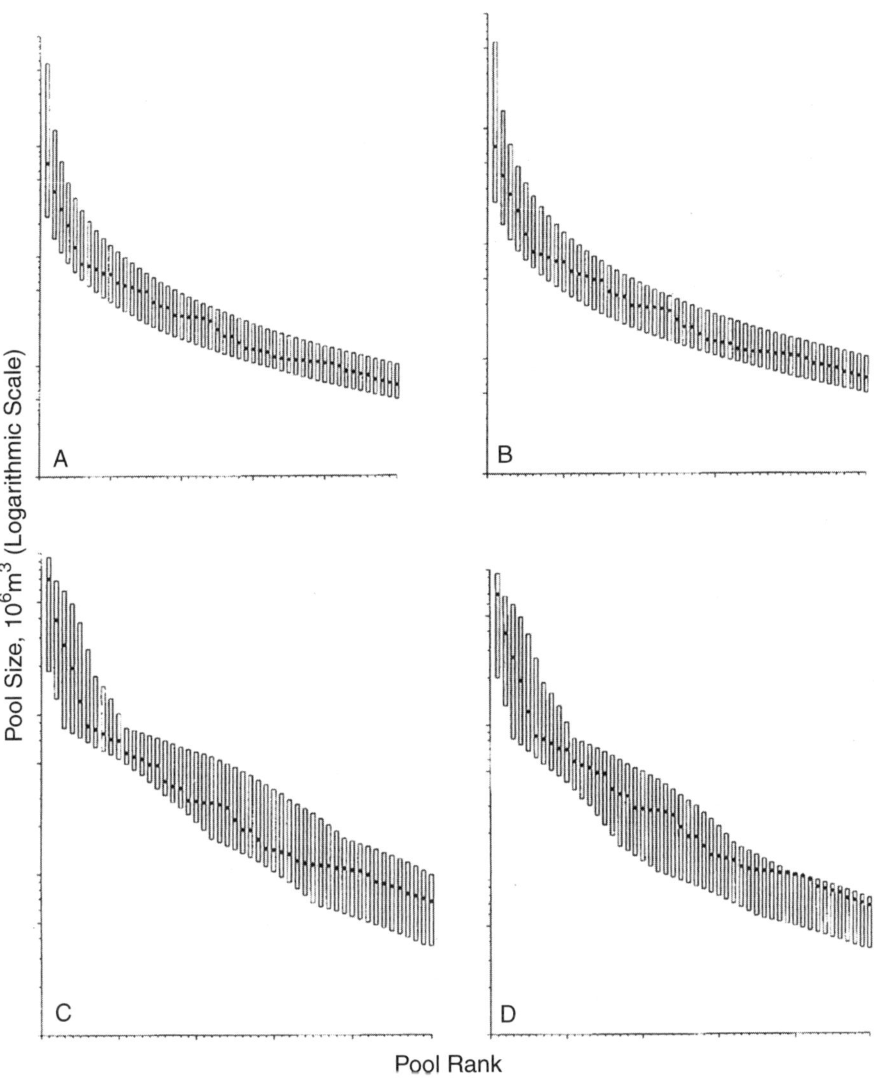

Figure 4.12. (A–D) Pool-size-by-rank plots for the mixed population of lognormal, Weibull, and Pareto populations derived by LDSCV when $n = 30$ (A) and $n = 50$ (B), and plots derived by NDSCV when $n = 30$ (C) and $n = 50$ (D). Prediction interval is the 0.9 probability level.

5. The 1000 pairs of estimated μ and σ^2 were plotted as box plots (Fig. 4.13). In these box plots, (a) the horizontal bar indicates the median value, (b) the box indicates the 50% intervals, and (c) the vertical bars represent the three standard deviations.

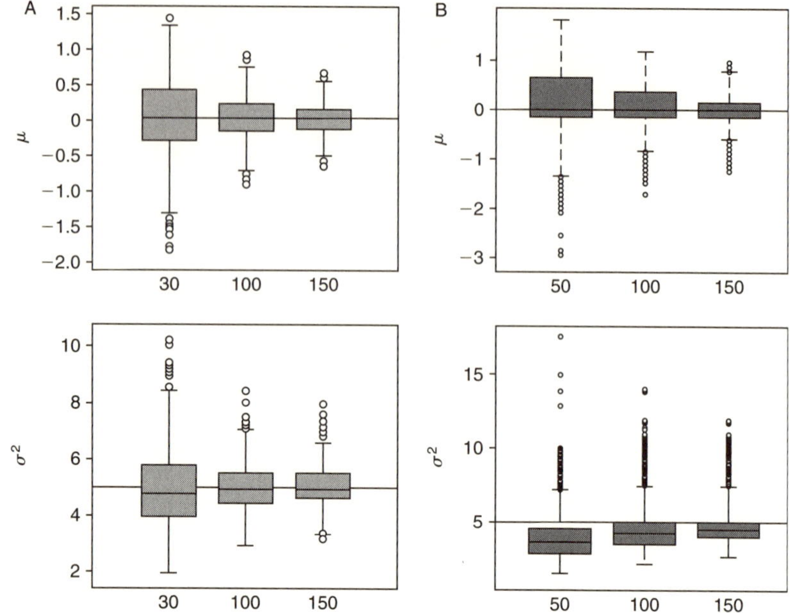

Figure 4.13. (A, B) Box plots displaying the estimates of μ and σ^2 derived by LDSCV (A) when sample size $n = 30$, 100, and 150; and NDSCV (B) when $n = 50$, 100, and 150. The 50% interval decreases as n increases.

The interpretation of the results is the following:

1. As the sample size n increases, the medians of μ and σ^2 approach the population values, and the 50% intervals of μ and σ^2 are reduced for both LDSCV and NDSCV cases.
2. The estimated values of μ and σ^2 fall into all 50% interval estimates, regardless of the sample size for the LDSCV case. For the NDSCV case, the value of μ does fall into the 50% interval, but σ^2 does not.

It can be concluded that the uncertainty is reduced when a sample size increases.

Validation by Retrospective Study

Jumping Pound Rundle Gas Play

Data on the gas reserves of the Jumping Pound Rundle pools booked in each year were collected, based on the past 40 years of records provided by the provincial government of Alberta, Canada. The play data set was divided into three time windows that were evaluated by LDSCV to compare the following estimates: (1) number of pools, (2) expected resource, (3) play resource distribution, and (4) sizes of the largest undiscovered pools. This approach allows us to examine the growth behavior of pool in-place booked reserves, as well as the appreciation and depreciation effects on petroleum resource evaluation results.

Figure 4.14A shows booked reserve variations in all pools belonging to the Jumping Pound Rundle gas pool growths from 1955 to 1993. The

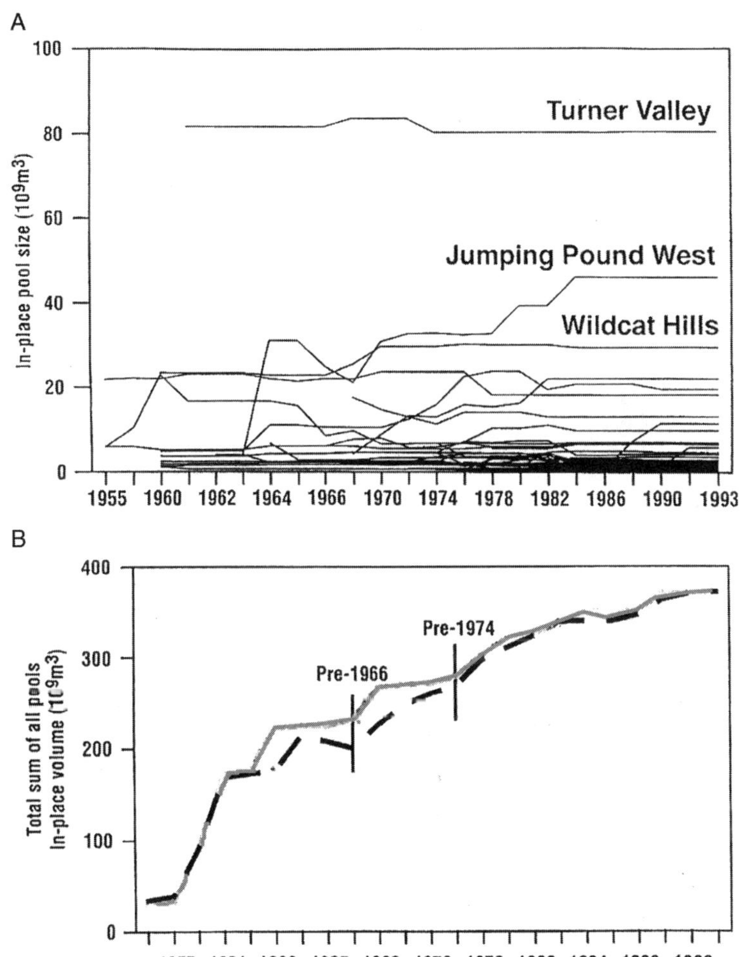

Figure 4.14. (A, B) Changes in reserve record from 1955 to 1993 for the Mississippian Jumping Pound Rundle gas play (Western Canada Sedimentary Basin). (A) Changes in reserves for each pool. (B) Cumulative changes in reserves booked by year (light line), and the booked reserve for 1993 (bold dashed line).

booked reserve of the Turner Valley pool did not substantially change, whereas the Jumping Pound West pool that had fluctuated increased significantly since its discovery. In general, fluctuations of booked reserves for the large pools are greater than those for small pools. From 1961 to 1980, the in-place booked reserves for the play were underestimated. Some of the booked reserves of the pools (e.g., the Jumping Pound West Rundle pool) have doubly appreciated since their discovery, whereas others have depreciated, but the total has only appreciated by a factor of 2% to 7%.

Figure 4.14B exhibits the Jumping Pound gas play growth for the same period. The dashed line shows changes of in-place volume recorded in 1993. The play data, comprising 94 discoveries as of 1991, were divided into three time windows: pre-1966 (Fig. 4.15A, left), pre-1974 (Fig. 4.15A, middle), and pre-1991 (Fig. 4.15A, right).

Assessment results are summarized in Table 4.5. The first column presents the time windows. In the second column, the total number of pools, N, is estimated using data from each time window. These values may be over- or underestimated. The third column records discovered reserves for specific years. The fourth column presents expected potential. The fifth column displays the total estimated play resource distribution (Fig. 4.15B). The 0.9 probability prediction intervals are of the same magnitude. The last column presents total play resource distribution means for the three windows, which are quite similar.

Individual pool sizes predicted from the pre-1966 time window of the Jumping Pound Rundle gas play are shown in Figure 4.16. The largest undiscovered pool, the Quirk Creek Rundle A, was discovered in 1967, whereas the second largest, the Clearwater Rundle A, was discovered in 1980. Although the pre-1966 time window predictions of the largest two pools are accurate, the entire prediction is not as good as those derived from the 1994 data set. Furthermore, no pool larger than the Clearwater Rundle A has been discovered since then.

Swan Hills Shelf Margin Gas and Leduc Isolated Reef Oil Plays

The Swan Hills shelf margin gas pools were also divided into two time windows: pre-1983 and pre-1994. Figure 4.17A displays the discovery sequences for the two windows. The play resource distributions for these two windows are shown in Figure 4.17B.

The Leduc isolated reef oil pools were divided into two time windows as well: pre-1965 and pre-1994. Their discovery sequences are shown in

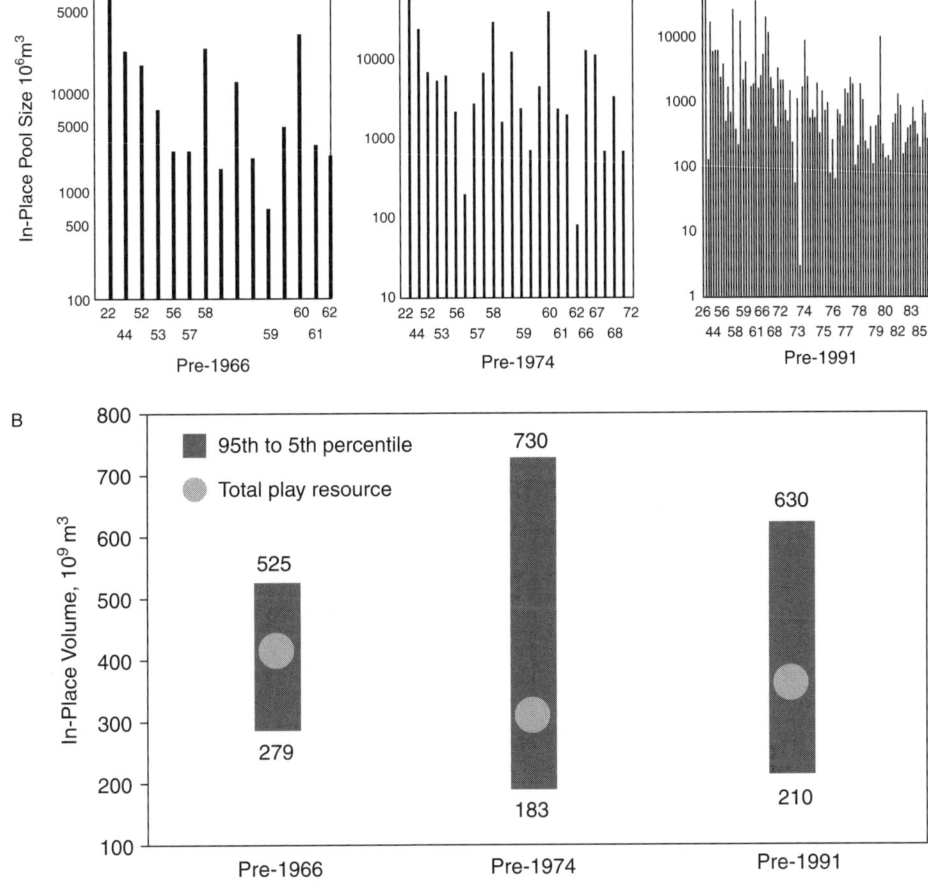

Figure 4.15. (A, B) Jumping Pound Rundle gas pools (A), divided into three time windows. (B) Play resource distributions for the pools from each time window after evaluation by LDSCV displayed as upper 95th percentile (lower end), 5th percentile (upper end), and the mean (circle). Data from the Western Canada Sedimentary Basin.

Figure 4.18A. The play resource distributions for the two windows are shown in Figure 4.18B. The 0.9 probability prediction intervals for the time windows of each play are similar.

Remarks

Assessment results are controlled by two factors: the quality of the pool reserves booked at the time of assessment and the number of discoveries available for the assessment. Gas and oil pool appreciation

Table 4.5. Summary of the Retrospective Study, Jumping Pound Gas Play

Time window	*Total no. of pools estimated*	*Discovered resource at time window*	*Expected potential*	*Play resource at 0.9 probability prediction interval*	*Expected resource*
Pre-1966	100	208	213	279–525	382
Pre-1974	100	262	63	183–730	376
Pre-1991	173	355	28	210–630	366

and/or depreciation in resource assessment results vary for different estimators. The results from the Jumping Pound example indicate that (1) the effect of reserve appreciation or depreciation on the estimation of play resource distribution is minimal, as shown in the comparison of play resource distributions for a specific time window (figs. 4.15B, 4.17B, and 4.18B); and (2) the expected potential decreases consistently as the amount of discovered resources increases—that is, the resources of the time windows are similar in each play.

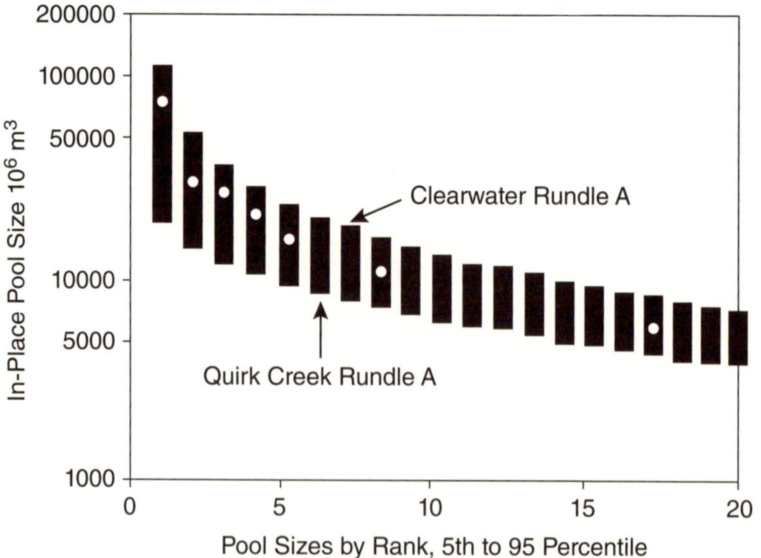

Figure 4.16. Pool-size-by-rank plot derived from a pre-1966 time window of the Jumping Pound Rundle gas play (Fig. 4.15). Note that the Clearwater Rundle A and Quirk Creek Rundle A pools were predicted.

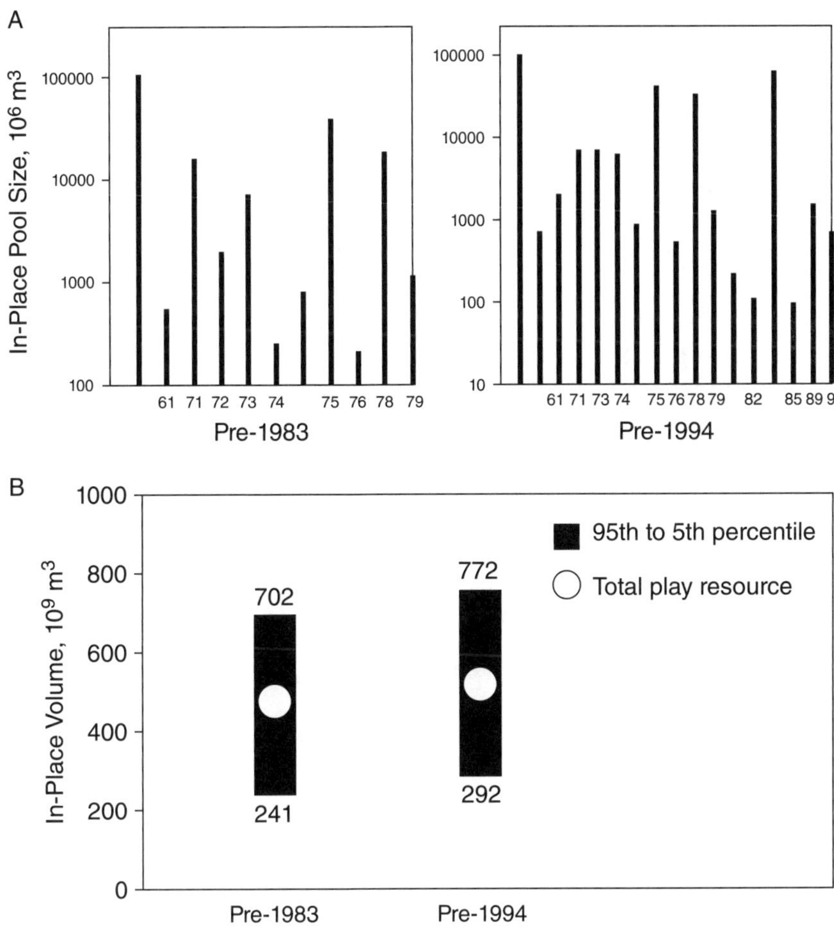

Figure 4.17. (A, B) Gas pools of the Devonian Swan Hills Shelf Margin play (A), divided into two time windows. (B) Play resource distributions after evaluation by LDSCV displayed as upper 95th percentile (lower end), 5th percentile (upper end), and the mean (circle). Data from the Western Canada Sedimentary Basin.

Impact of Nonproductive and Noncommercial Pools

The areal extent of a play does not directly influence the pool-size distribution derived from the two discovery process methods LDSCV and NDSCV. In cases when the additional area includes wildcats with substantial drill stem test recoveries that might become pools with additional development, the play area might increase because more pools are included. Therefore, play area is not a direct factor influencing pool-size distribution. In this aspect, the models differ from the methods of Arps and Roberts (1958), and Drew (1990; Drew et al., 1980, 1982).

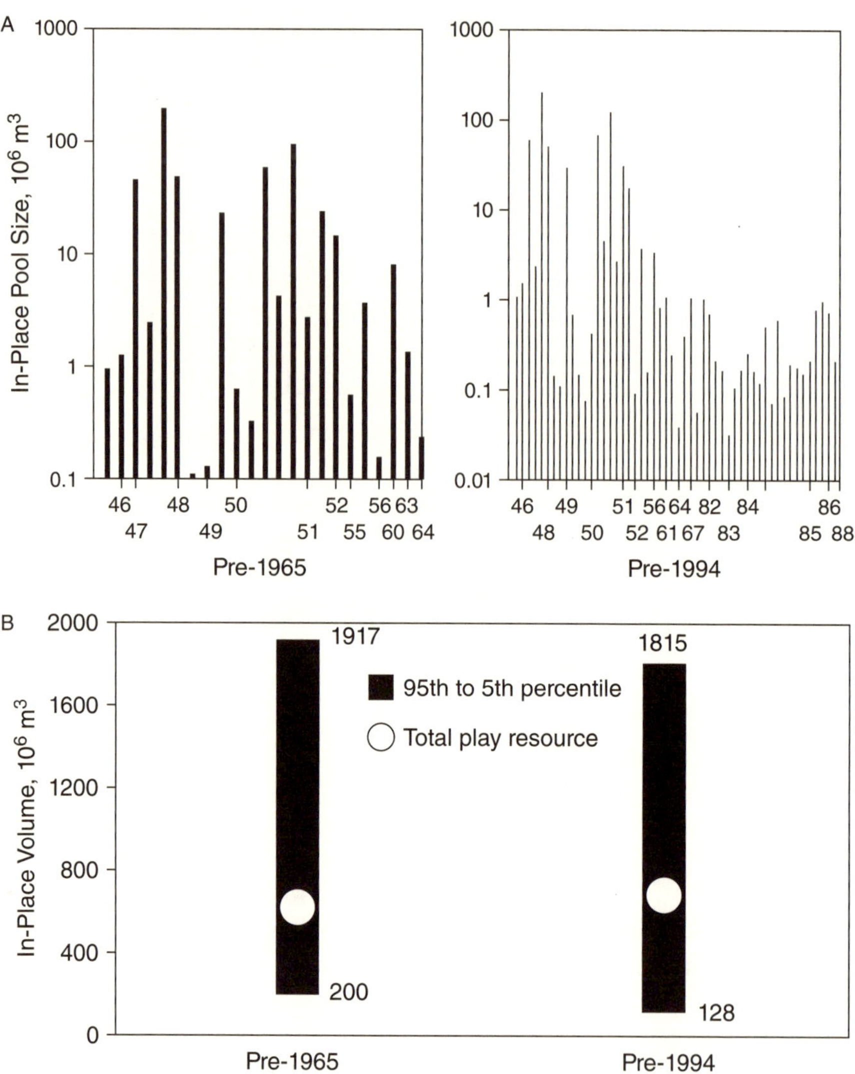

Figure 4.18. (A, B) Oil pools of the Devonian Leduc isolated reef play (A), divided into two time windows. (B) Play resource distributions after evaluation by LDSCV displayed as upper 95th percentiles (lower end), 5th percentile (upper end), and the mean (circle). Data from the Western Canada Sedimentary Basin.

Impact of a Nonproductive Trap

Another question that has been raised is: How do the methods handle a case when pools have been deposited, but were subsequently destroyed by geological processes before exploration began? In this case will the discovery process models predict the pools destroyed by nature? The answer is no, for the following reason. Discovery process models are controlled not only by physical features of deposition, but also by the manner in which the pools are discovered. Therefore, if the pools have been destroyed by geological processes, the discovery probabilities will not be the same because the pools destroyed by nature are excluded from the maximum-likelihood function (see Eq. 3.5).

Take a simple case, for example, where $N = 5$ with pool sizes x_1, x_2, x_3, x_4, and x_5. The probability of discovering pool x_1 is $x_1/(x_1 + x_2 + x_3 + x_4 + x_5)$. Suppose x_2 has been destroyed by geological processes, then the probability of discovering pool x_1 is $x_1/(x_1 + x_3 + x_4 + x_5)$. Note that x_2 is not included in the probability statement. This trivial example demonstrates that nonproductive traps have no impact on the estimates derived from discovery process models.

To illustrate the concept of a discovery process model, let us do some "marble fishing." Suppose we have an urn that contains red, green, and black marbles of different sizes. The red and green marbles represent gas (red) and oil (green) pools, and the black marbles represent nonproductive traps. When fishing for marbles, we cannot see or touch them, but intend to "fish out" large red or green marbles and not the black ones. What we know is that our marble-fishing technique allows us to fish out large red or green marbles with high probability. Whenever we obtain either a red or green marble, we place it in a time sequence. The black marbles we fish out are discarded as failed prospects. All the marbles we have fished out are not put back in the urn. This is the statistical procedure of sampling without replacement. After several trials, the following questions arise:

- From the time sequence, can we estimate how many red and green marbles are yet to be fished out?
- What are the sizes of the remaining red and green marbles?

The discovery process model is designed to estimate the number and size of the remaining red and green marbles (not the black ones).

Impact of Missing Pools

Commonly, small pools are not reported because of economic truncation. The absence of small pools from discovery sequences raises a problem in petroleum assessment. To investigate the impact of their absence from a discovery sequence, the lognormal population (Fig. 4.3) is used. Suppose that pools with sizes less than 0.5×10^6 m^3 were not reported. After this economic truncation process, the final number of pools equals 183 instead of 300.

The resultant sequence was subjected to analysis by LDSCV and NDSCV. The relationships between the values of N and the log likelihood are summarized as follows. In the case of missing small pools, LDSCV predicts that the value of $N = 200$, whereas NDSCV plateaus at 220. Although both models do not predict the truncated value of N exactly, the estimated N's are close to the true value, 183. If more small pools were missed from the discovery sequence, then the log L versus N relationship would degenerate into either a negative exponential or no pattern at all. This confirms that the missing pools affect the quality of the assessments.

Testing the Adequacy of Probability Distributions

Essentially, probabilistic statistical analysis is based on the assumption that a set of data arises as a sample from some class of probability distribution. Tests of distributional assumptions have been an important subject in petroleum resource evaluation procedures. Kaufman (1965) used a lognormal pool-size distribution to describe oil and gas pools. McCrossan (1969) plotted the discovered oil and gas pools from the Western Canada Sedimentary Basin on logarithmic probability paper and found that the plots tended to be straight lines. Power (1992) applied the Anderson–Darling test to several plays assessed by Podruski et al. (1988) from the Western Canada Sedimentary Basin and concluded that they follow a lognormal distribution, whereas others follow a Weibull distribution. The statistical assumption of this test is that all oil pools are randomly discovered by geologists. This assumption is incorrect. The test of a data set, which is a biased population sample, is an unsolved problem.

This section attempts to solve this problem and presents an informal quantile–quantile (Q–Q) plot to assess distributional assumptions. The information required is based on the results of the nonparametric estimates, $\hat{p}_i$ (refer to "Nonparametric Discovery Process Model" in Chapter 3). The advantage of the procedure is that it is not based on any assumption about the shape of a probability distribution. However, the procedure assigns mass only to the observed data and assumes that the largest pool in the population is no larger than the largest pool in the sample, and that the smallest undiscovered pool is no smaller than the smallest discovered one. This is an unrealistic situation. To overcome this disadvantage, the $\hat{F}$ estimated is approximated by various probability distributions. Then the best fit among the distributions is judged using the informal graphic procedure.

The Procedure

Suppose that $\hat{F}$ is an estimate and is being tested to determine whether it is equal to a hypothesized distribution F_0. A number of graphic methods can be applied to test the hypothesis. The percent–percent (P–P) plot is checked to determine whether it falls along a straight line through the origin with a slope of one. However, the P–P plot has two disadvantages. First, it only allows one to check the adequacy of completely specified distributions. In practice, it would be used more to determine the shape of the distribution, such as lognormality. Second, if the plot is nonlinear, it becomes difficult to determine which alternative shapes one should consider.

The Q–Q plot, on the other hand, is designed to overcome the drawbacks inherited from P–P plots and can be used to assess the adequacy of a hypothesis whether a data set comes from a family $F_0[y-\mu/\sigma]$ for an unknown location parameter μ and scale σ^2. If we consider that the data set is from a distribution with shape F_0, the data will follow a linear configuration. So one needs only look for linearity without having to estimate values for μ and σ^2. If linearity does exist, then the intercept of the line is an estimation of μ, and the slope is an estimation of σ^2. Departures from the straight line in the theoretical Q–Q plot clearly indicate that the observed and theoretical distributions do not match. When data points do not show a straight line on a plot, then they may indicate the nature of the mismatch, such as (1) presence of outliers at either end; (2) curvature at both ends, indicating long or short tails at both ends; (3) convex or concave curvature, related to symmetry; and (4) plateaus. The significance of these mismatches (Chambers et al., 1983) will be discussed later.

Interpretation

Outliers

Samples of geological populations often contain outliers. When they are encountered in a set of data, it is prudent to examine the source of the data, if possible, to verify the values. If the values are in error, they can be corrected or set aside, but if they really belong to the population, they might be the most important observation in the sample.

Long or Short Tails at Both Ends

Another departure from linearity often observed in Q–Q plots is long or short tails. The ends of the configuration curve up to the right and down to the left. A straight line can be fitted to the center portion of the plot. This indicates that these data represent longer tails on the right than the hypothesized distribution F_0 (Fig. 4.19A).

Symmetry

If the Q–Q plot forms an S shape (Fig. 4.19B), then the data have a shorter (lighter) tail than that of the hypothesized distribution F_0.

Plateaus

Distinct clusters of points that are not accounted for by the theoretical distribution are referred to as *plateaus.* Currently, PETRIMES provides 12 types of probability distributions for testing the adequacy of statistical assumptions: normal, half-normal, uniform, gamma, lognormal, power normal, Pareto, shifted Pareto, truncated

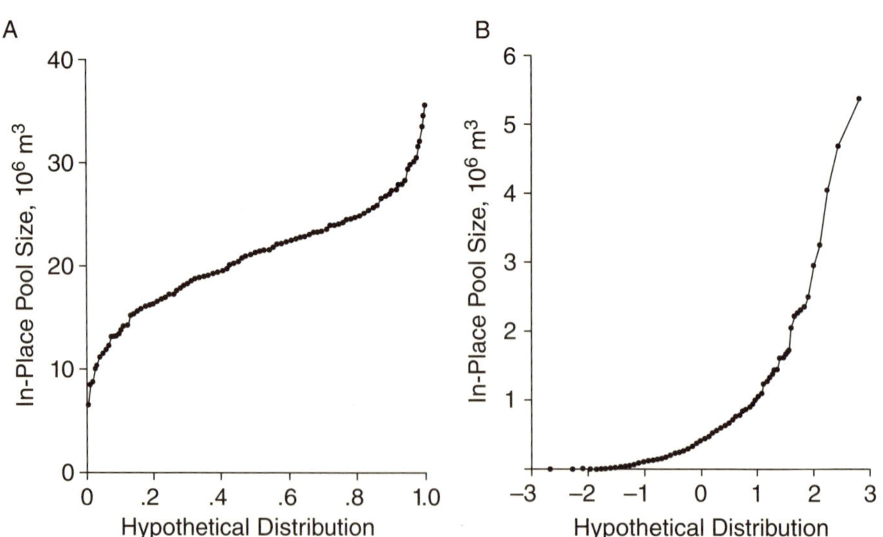

Figure 4.19. (A, B) Example theoretical Q–Q plots. (A) The plot is concave up at the right side and convex down at the left side, implying that the right-side tail of the hypothetical curve is shorter than that of the data. (B) The plot is an S-shaped curve, implying that the right tail of the hypothetical distribution is shorter than that of the data.

and shifted Pareto, Weibull, one-parameter exponential, and two-parameter exponential.

The Beaverhill Lake Play

For the Beaverhill Lake play, all distributions were hypothesized and fitted to the nonparametric equation as displayed previously in Figure 3.6, line B. The Q–Q plot results for these tests are displayed in figures 4.20 and 4.21. The assessment of the distributional assumption is summarized as follows:

1. The Beaverhill Lake data set has a longer tail than that of the distributions of normal, power normal (with power = 0.5), uniform, gamma (with shape factor = 5 to 0.01), one-parameter exponential, and two-parameter exponential.
2. The Q–Q plots for the truncated and shifted Pareto distribution display an S shape. This means that the tail of the distribution is longer than that of the data set.
3. The lognormal, Weibull, and power normal (with power = 0.001) distributions might have a slightly longer tail than that of the data set. However, the lognormal is a better choice if one has to use a prior distribution. Statistics for the straight line fitted to various distributions are listed in Table 4.6. From this, one can judge which one or two distributions are better for a specific play. The nonparametric discrete distribution of Figure 3.6, line B is approximated by a continuous lognormal distribution that is used to estimate individual pool sizes.

Furthermore, PETRIMES can estimate the ratio from the empirical pool-size distribution by computing the shape factor, θ, value. An example is presented by Lee and Gill (1999).

Plays from Worldwide Basins

The Western Canada Sedimentary Basin provides a vast and valuable information source to test the adequacy of probability distributions. Gas plays that have been tested by the nonparametric model include the Devonian (Reinson et al., 1993), Mississippian and Permian (Barclay et al., 1997), Triassic (Bird et al., 1994), and Foothills (Lee et al., 1995; Osadetz et al., 1995). More data sets obtained from worldwide basins

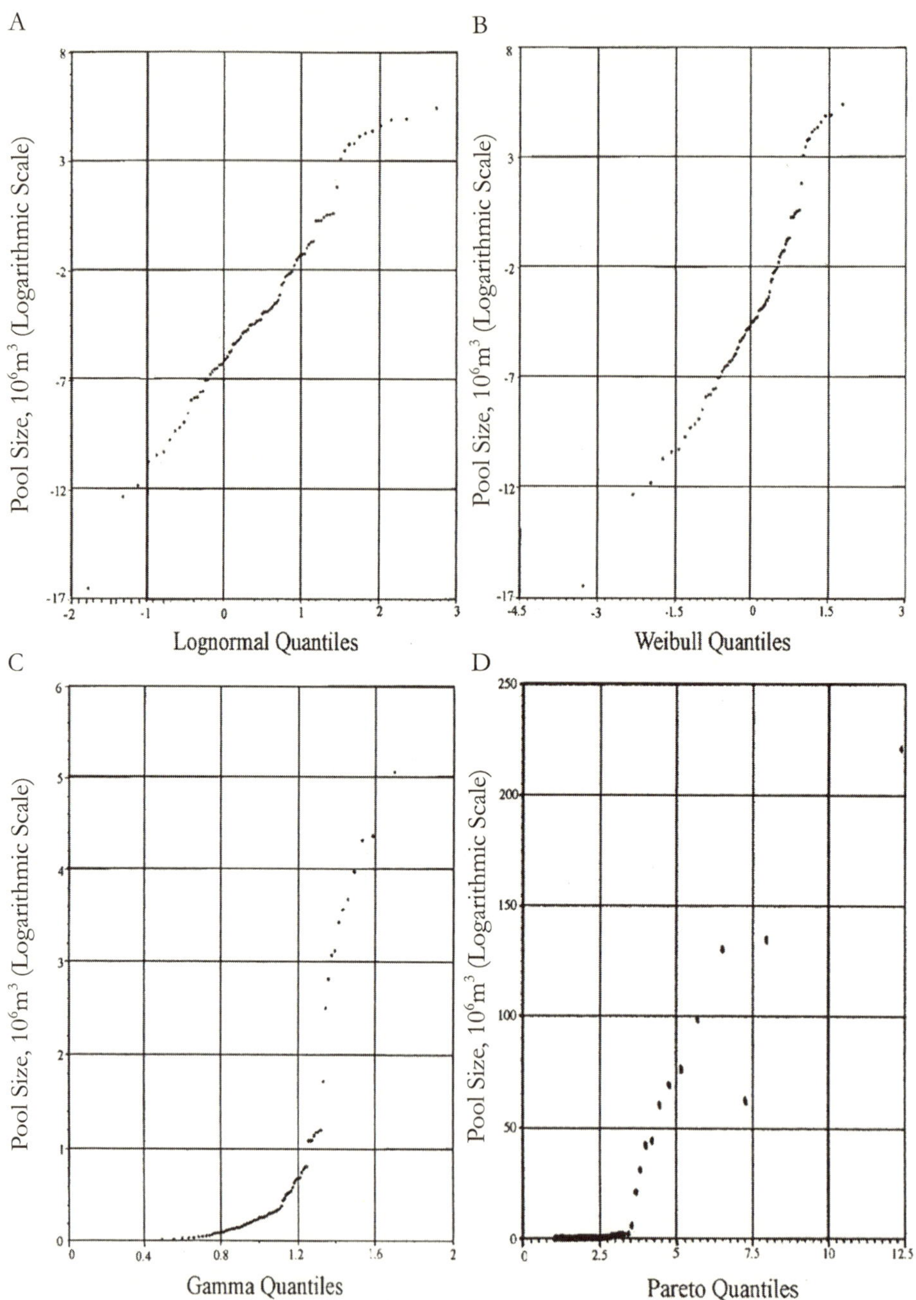

Figure 4.20. (A–D) Q–Q plots for Devonian Beaverhill Lake reef play. The plots indicate that the lognormal distribution (A) is not perfect, but is the best among the four distributions tested. Data from the Western Canada Sedimentary Basin.

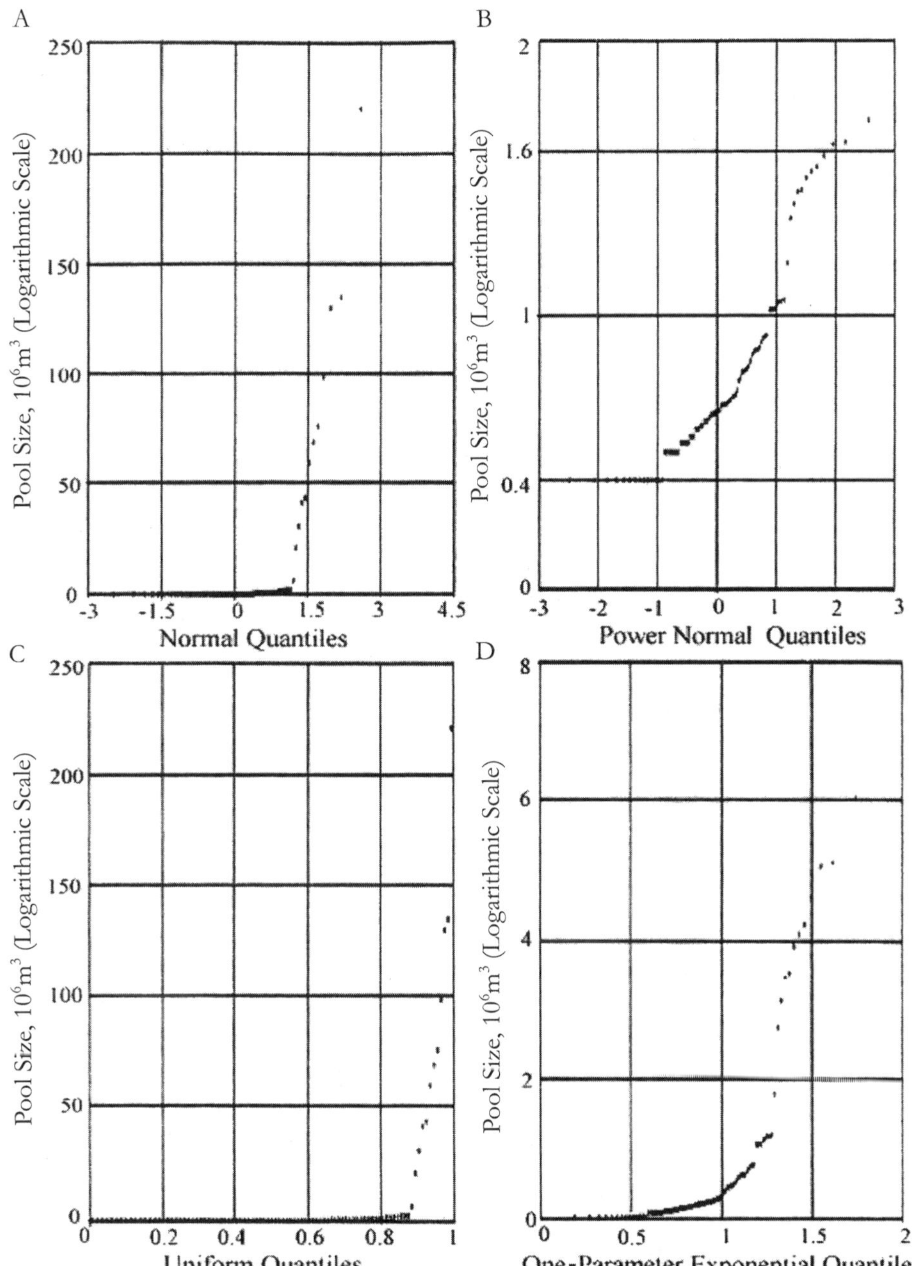

Figure 4.21. (A–D) Additional Q–Q plots for the Devonian Beaverhill Lake reef play. Plots indicate that the right-side tails of the normal (A), uniform (C), and one-parameter exponential (D) distributions are too short for the play data set. Data from the Western Canada Sedimentary Basin.

Table 4.6. Statistical Parameters for Various Probability Distributions of the Beaverhill Lake Play

Probability distribution	*Intercept, a*	*Slope, b*	*Correlation coefficient, r*	*Standard error*
Half normal	–0.091	0.017	0.590	0.617
Normal	0.701	0.013	0.710	0.179
Power normal	–1.956	2.653	0.926	0.135
Lognormal	0.901	2.358	0.972	0.052
Weibull	0.485	0.274	0.906	0.260
Uniform	0.484	0.004	0.474	0.062
Gamma	0.764	0.203	0.771	0.035
One-parameter exponential	0.768	0.187	0.780	0.040
Two-parameter exponential	0.808	0.024	0.823	0.319
Truncated and shifted Pareto	0.392	0.029	0.945	0.111

have also been tested using Q–Q plots. Some of the outliers contained in the samples might not follow straight lines. However if the outliers are excluded from the Q–Q plots, one can make the following conclusions from observations of more than 100 plays:

1. In all cases, lognormal distributions are the most appropriate distributions for the plays tested, as shown in figures 4.22 and 4.23.
2. Generally, the Weibull distribution exhibits a concave curve in the Q–Q plots. Figure 4.24 shows that it is the best distribution for this play. This is one of only two plays from more than 100 plays studied for which the Weibull distribution is best. However, the lognormal or power normal distributions are also appropriate.
3. In the Pareto Q–Q plots, all the play data sets are compressed into a small area in the lower left end of the plot. An exception is that presented in Figure 4.25, which shows that the Pareto distribution is the best of the four for this play. This is the only play from more than 100 plays studied for which the Pareto distribution is best. The Pareto distribution may sometimes be adequate for the largest few pools (Fig. 4.26).

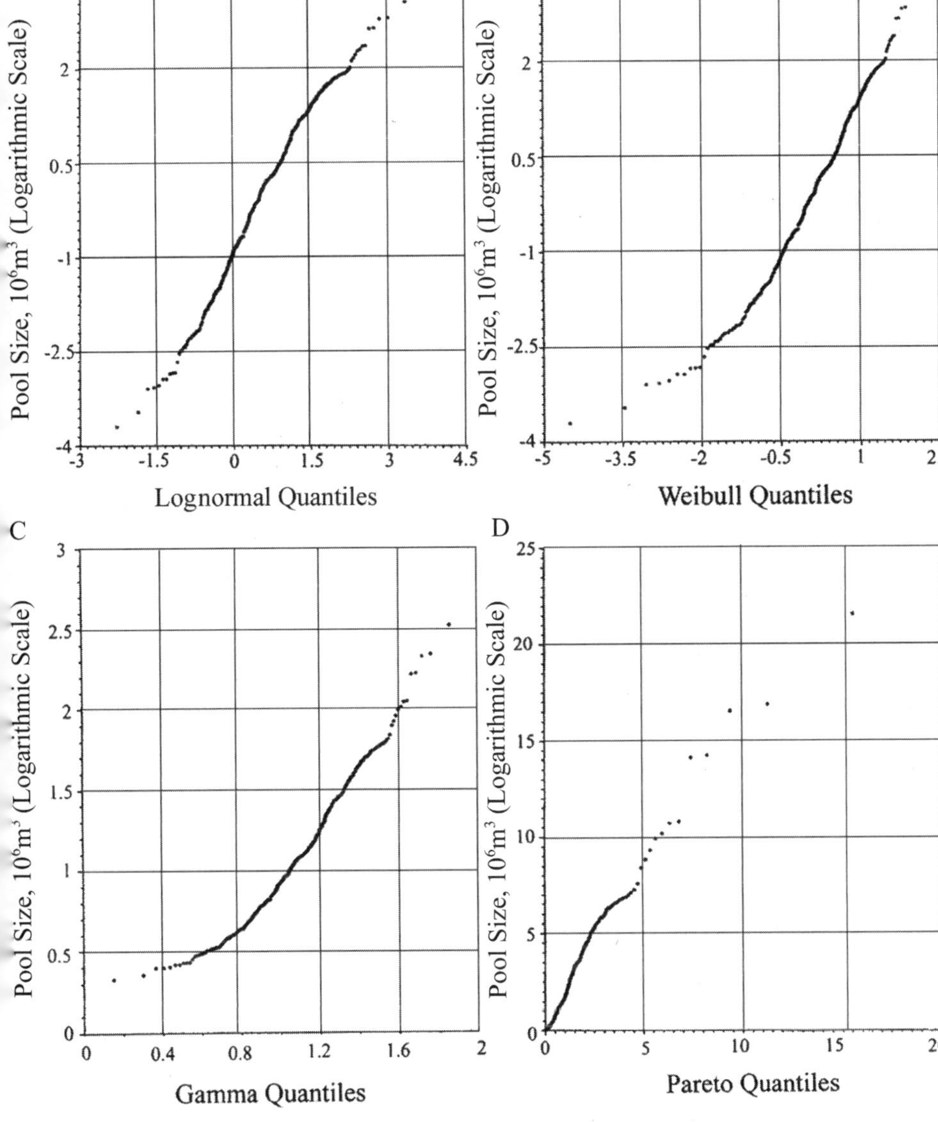

Figure 4.22. (A–D) Q–Q plots for the Middle Silurian Niagaran pinnacle reef play, northern Michigan, USA. Plots indicate that the lognormal distribution (A) is the best among these four distributions. Data from Gill (1994).

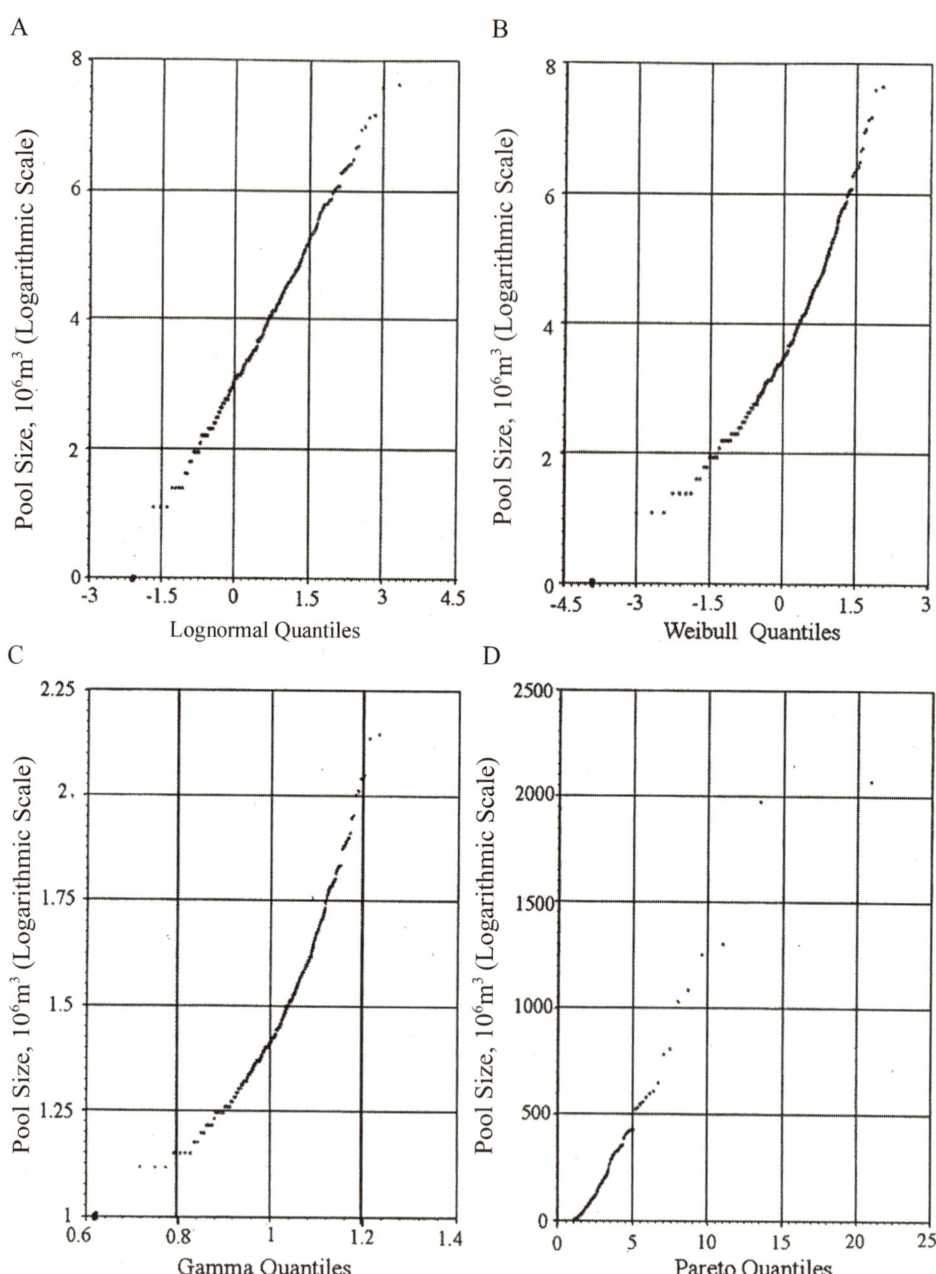

Figure 4.23. (A–D) Q–Q plots for the Cretaceous Glauconitic sandstone play. Plots show that the lognormal distribution (A) is the best choice among the four distributions tested. Data from the Western Canada Sedimentary Basin.

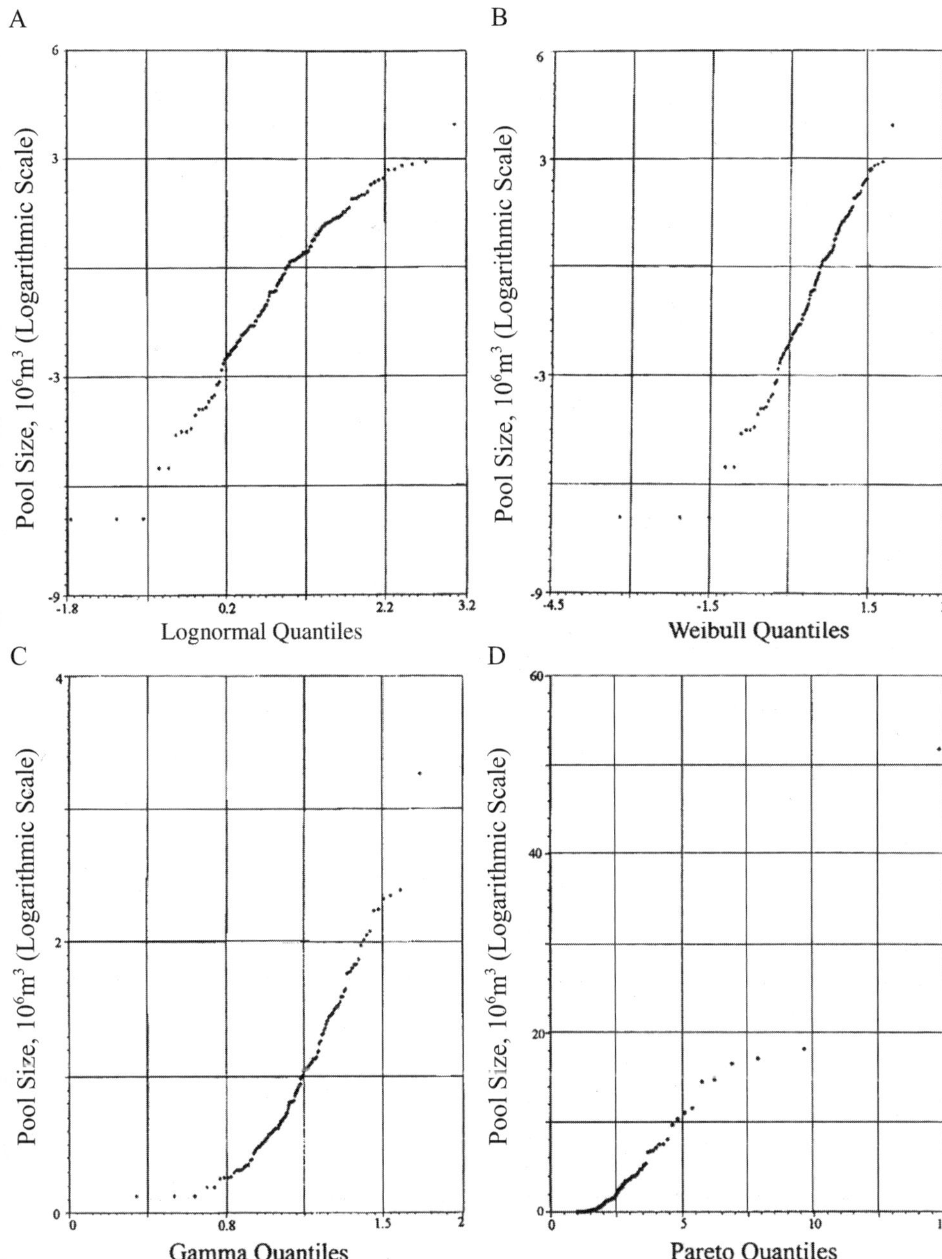

Figure 4.24. (A–D) Q–Q plots for the Minnelusa play, Powder River Basin, Montana–Wyoming, USA. Plots show that the Weibull distribution (B) is the best for this play. This is one of only two populations from more than 100 plays studied for which the Weibull distribution is best.

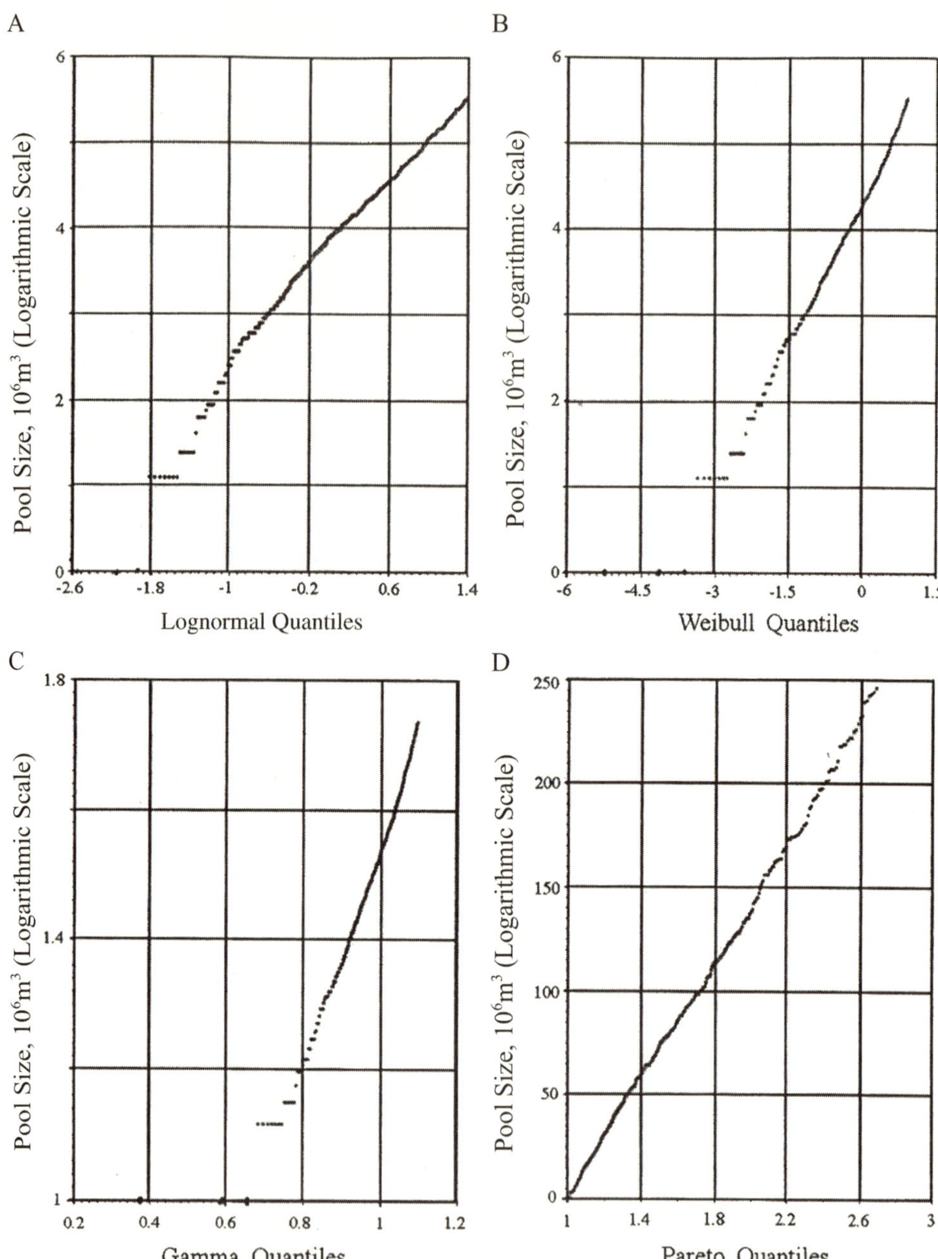

Figure 4.25. (A–D) Q–Q plots for the Cretaceous Gething/Dunlevy play, Deep Basin. Plots show that the Pareto distribution (D) is the best for this play. This is the only play from more than 100 plays studied for which the Pareto distribution is best. Data from the Western Canada Sedimentary Basin.

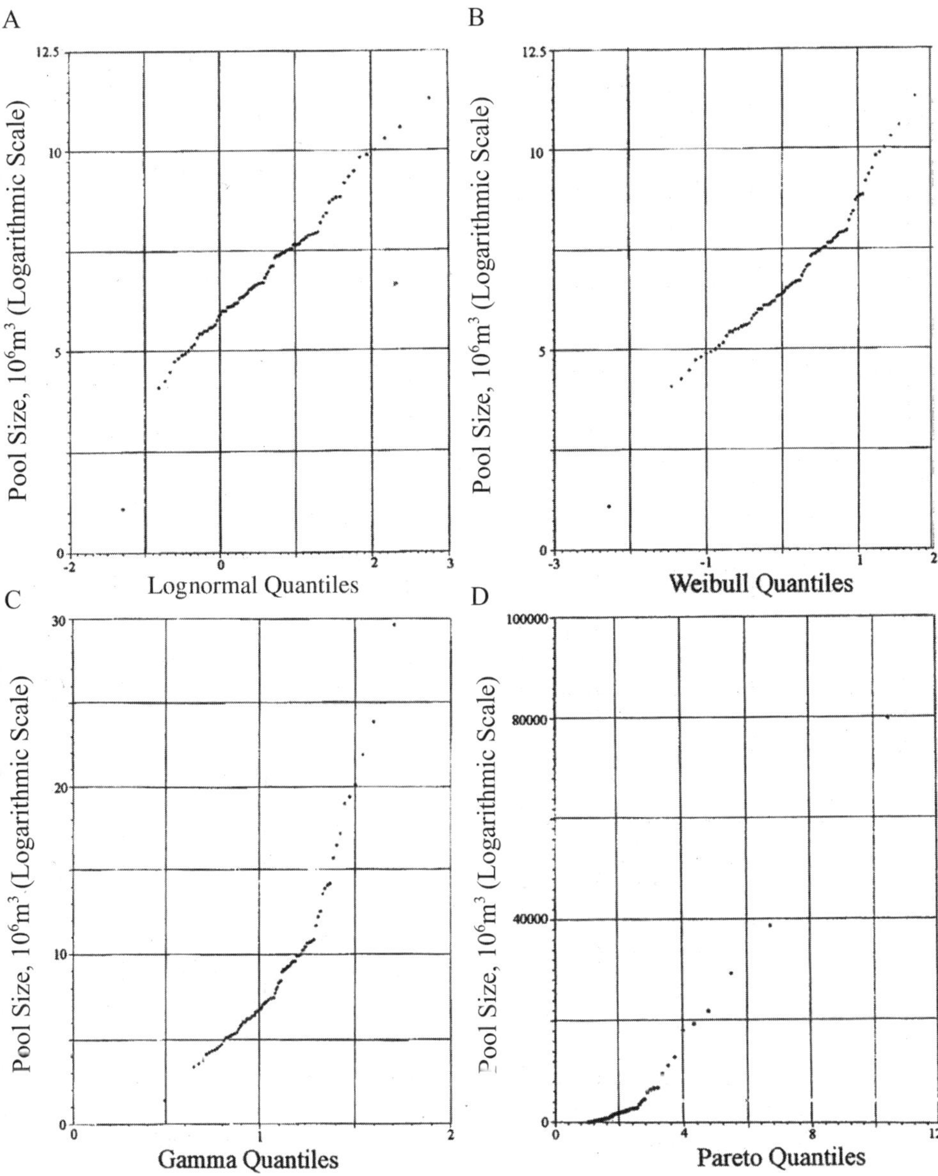

Figure 4.26. (A–D) Q–Q plots for the Mississippian Jumping Pound Rundle gas play, McConnell Thrust. Plots show that the largest few pools might be approximated by a Pareto distribution (D). Data from the Western Canada Sedimentary Basin.

Pool-Size Distribution of a Basin

Observations from 25 Devonian plays (Reinson et al., 1993) with different means and variances of lognormal and empirical distributions have been considered as a single population. For the superpopulation case, each play was evaluated using LDSCV or NDSCV. For the finite population case (the maximum-likelihood function of the finite population approach is based on the work by Bickel et al. [1992]), all pool sizes were combined into one single population with a size class of base 2 (Table 4.7). Conclusions are as follows:

1. The grouped distributions for both the superpopulation and the finite population approaches exhibit a J shape on a logarithmic scale.
2. If the ratios of the two largest and smallest classes are ignored (Table 4.8), we then obtain
   - the ratio derived from the superpopulation approach ranging from 1.0 to 2.3, with a mean of 1.6
   - the ratio derived from the lower limit of the finite population approach ranging from 1.0 to 2.3, with a mean of 1.7

Table 4.7. Comparisons between the Estimates Derived from the Superpopulation and Finite Populations When the 25 Devonian Mature Gas Plays Are Considered as a Single Population

Pool size class, 10^6 m^3	*No. of discovered pools*	*Predicted undiscovered pools*		
		Superpopulation	*Finite population*	
			Lower limit	*Upper limit*
<64	1158	6500	6657.5	57,812
64–128	239	643	653	4995
128–256	191	404	299	2335
256–512	155	219	144	1231
512–1024	98	125	53	532
1024–2048	54	59	12	159
2048–4096	33	21	2	42
4096–8192	25	7	0	10
8192–16,384	21	0	0	1
16,384–32,768	0	0	0	0
32,768–65,536	10	0	0	0
>65,536	1	0	0	0

Table 4.8. Ratios between Two Adjacent Pool Size Classes of Table 4.7

Pool size class, 10^6 m^3	*No. of discovered pools*	*Ratios*		
		Superpopulation	*Finite population*	
			Lower limit	*Upper limit*
<64	4.9	8.7	25.9	11.3
64–128	1.3	1.8	1.8	2.1
128–256	1.3	1.8	1.7	1.8
256–512	1.6	1.6	2.0	2.2
512–1024	1.8	2.0	2.3	3.0
1024–2048	1.6	2.3	1.9	2.8
2048–4096	1.3	1.9	1.4	2.1
4096–8192	1.2	1.4	1.2	1.6
8192–16,384	2.1	1.2	2.1	2.1
16,384–32,768	1.0	1.0	1.0	1.0
32,768–65,536	10.0	10.0	10.0	10.0
>65,536	—	—	—	—

- the ratio derived by the upper limit of the finite population approach ranging from 1 to 3 , with a mean of 2
- the sample ratio ranging from 1.0 to 2.1, with a mean of 1.4

The sample ratio is smaller than that of the population. The finite population approach, which does not require any prior probability distribution, produces a more irregular ratio.

This example leads to the following discussions. A natural basin population, which consists of a mixture of several lognormal and empirical distributions, can form a J-shaped distribution. From the examples studied, there is no apparent trend for all ratios. Does the absence of a trend imply a constant ratio? Is it possible that the ratio varies from class to class without any pattern? Should we consider these variations random phenomena that can be represented by their means? Or are these variations natural anomalies? In these cases, the number of pools would be under- or overestimated if an average ratio or any ratio were used to predict the entire population. Therefore, the hypothesis that there is a constant ratio between two size classes remains unproved.

The previous discussion suggests that a J-shaped distribution, either directly observed or statistically derived from a sample, does not necessarily indicate that its superpopulation distribution belongs to a Pareto distribution family.

Justifications for Using a Lognormal Distribution

To this point we have examined Q–Q plots from worldwide examples. We shall now choose a specific distribution from a lognormal family to represent a play or geological population and discuss the following topics:

- Evidence from the Q–Q plots
- Approximation of a lognormal distribution to geological random variables
- Advantages of using a lognormal distribution
- Estimation error resulting from lognormal distribution approximation

Evidence from the Q–Q Plots

From the examples studied and the Q–Q plots constructed from the output of the nonparametric estimation, we observed that the lognormal distribution was and still is a favorable choice among the distributions tested. The Weibull and gamma distributions usually displayed a concave upward pattern in their Q–Q plots for the plays studied. This concave upward feature implied that the right-hand tail of the gamma and Weibull distributions (large size) was too short for the play data sets tested. On the other hand, the truncated and shifted Pareto distributions exhibit an S-shaped pattern in their Q–Q plots. These patterns implied that the right-hand tail was too long for the play data sets. For prediction of the largest pool size in the population, the truncated and shifted Pareto distribution would tend to yield a much larger pool. Similar results were obtained by Houghton (1988) and Davis and Chang (1989). If a distribution tail were too long or too short, then the total resource of a play would be over- or underestimated respectively. In most cases, the Q–Q plots of lognormal distributions are almost straight lines.

Approximation of a Lognormal Distribution to Geological Random Variables

Examples from the Western Canada Sedimentary Basin demonstrate that a lognormal distribution is adequate for approximations of various large sample sets. Take the data sets from some mature plays in the Western Canada Sedimentary Basin (Fig. 4.27), for instance. The pool area of the Cardium sandstone (Fig. 4.27A), porosity (Fig. 4.27B), and net pay (Fig. 4.27C) of the Lower Mannville sandstone, and the net pay of the Devonian clastics (Fig 4.27D) can be approximated by the families of the lognormal distribution. Figure 4.27A displays a peculiar pattern. The large steps between 60 and 100 ha are the result of the assignment of 64 ha to some of the small pools. In these cases, a prior distribution such as the lognormal can provide a framework for estimating the population distribution.

If a pool-size distribution is computed from the products and divisions of several dependent or independent lognormal distributions, then the end product is lognormal. According to the central limit theorem, the end product also tends to be a normal or lognormal distribution, regardless of the original probability distribution types. These probability distributions can be area of pool, net pay, formation thickness, porosity, water saturation, and others.

Oil and gas pools form as the result of the following processes. First, organic matter is deposited in a bed to form the source rock, after which it is transformed into oil and/or gas when the source rock is buried deep enough to generate oil or gas. Oil and gas migrate from the source rock and are trapped in the final reservoir. Countless minute oil drops and gas bubbles accumulate in tiny traps and may leak to the surface as seepage or gas bubbles. If we use a probability distribution to express the quantities of the result of each process, then the end product of all geological processes can be equivalent to the multiplication of these distributions together as a single distribution. The law of proportionate effect (Aitchison and Brown, 1969, pp. 22–23) supports the deduction that the end products of the geological processes, oil and/or gas pools, are lognormally distributed.

Advantages of Using a Lognormal Distribution

For the immature or conceptual plays, probability distributions for all geological variables are constructed from interpretations of geological information. The distributions constructed reflect current knowledge. In these cases, if the assumption of lognormality is used, then geologists will be able to examine the sizes of the largest few pools without predetermination by assessors. Lognormal distributions adequately approximate the distributions recognized by geologists (Lee and Wang, 1983b).

In addition, correlation among variables can be conveniently handled with a lognormal distribution. Refer to the section on lognormal approximation in Chapter 5 for details.

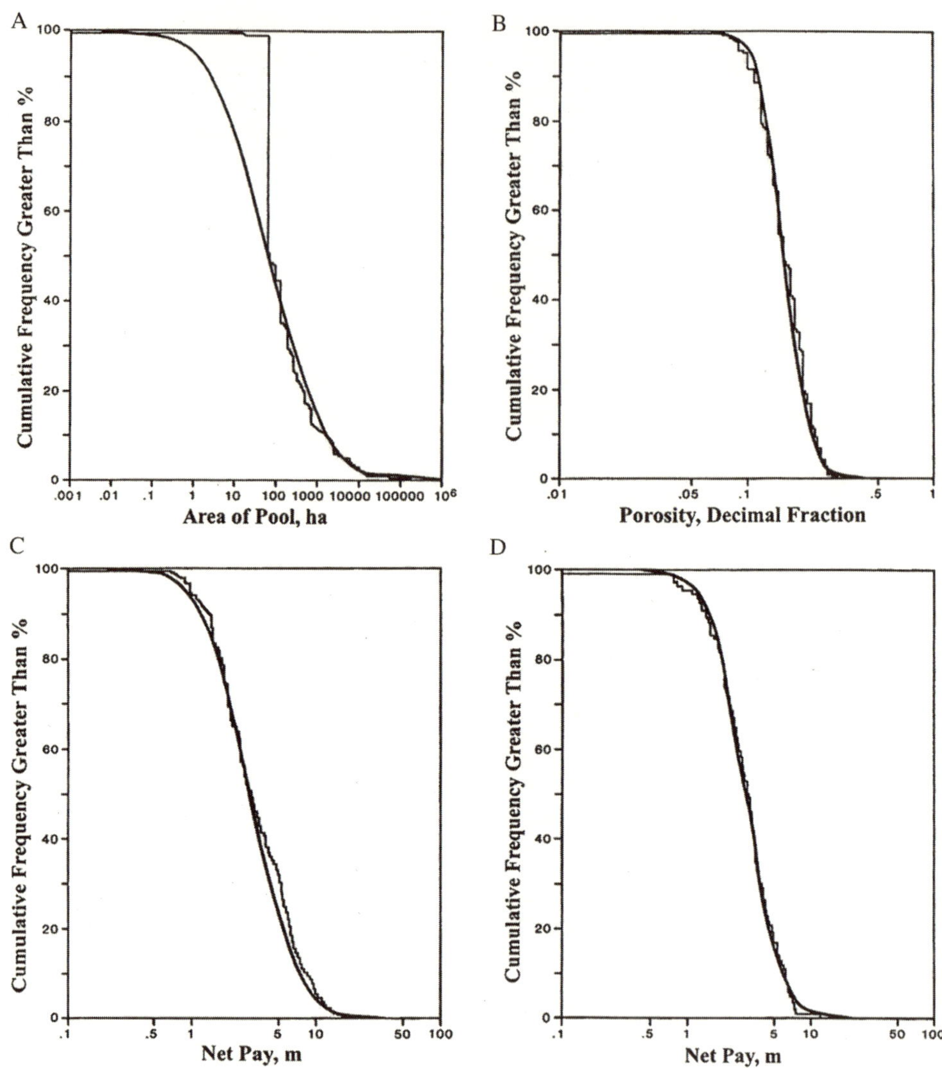

Figure 4.27. (A–D) Cumulative greater-than plots showing various geological random variables. Pool area of Cardium sandstone (A), porosity of Lower Mannville sandstone (B), net pay of Lower Mannville sandstone (C), and net pay of Devonian clastics (D) from the Western Canada Sedimentary Basin can be approximated by the family of a lognormal distribution.

The Pareto model may provide estimates about the small pools but requires that pool size exhibit a constant ratio of two adjacent size classes. Drew et al. (1980) empirically estimated that the ratio was about two for the Permian Basin. Chen (1993) plotted log density versus log pool size, and all the pools beyond the first break on the right-hand side of the plot had already been discovered. Therefore, the average ratio or a single ratio obtained from discovered pools can be applied to estimate the number of small-size pools. These two methods are ad hoc procedures. Furthermore, a J-shaped sample distribution is not indicative of a Pareto population (Lee, 1993b).

Estimation Error Resulting from Lognormal Distribution Approximation

If we apply a lognormal assumption to the Weibull, gamma, and Pareto distributions, the errors resulting from an incorrect statistical assumption decrease as the sample size increases (see tables 4.1 through 4.4).

The lognormal family (including the power normal) is not perfect, but has proved to be the best among the four distributions tested. It can, at least, describe the economically viable portion of the play and yield estimates for ongoing exploration planning.

5

Evaluating Conceptual Plays

As time goes on, qualitative methods are replaced by quantitative methods.

—F. Y. Loewinson-Lessing

A *conceptual play* has not yet been proved through exploration and can only be postulated from geological information. An *immature play* contains several discoveries, but not enough for discovery process models (described in Chapter 3) to be applied. The amount of data available for evaluating a conceptual play can be highly variable. Therefore, the evaluation methods used are related to the amount and types of data available, some of which are listed in Table 5.1.

Detailed descriptions of these methods are beyond the scope of this book. However, an overview of these and other methods will be presented in Chapter 7. This chapter deals with the application of numerical methods to conceptual or immature plays. For immature plays, discoveries can be used to validate the estimates obtained. In this chapter, the Beaverhill Lake play and a play from the East Coast of Canada are examined.

Geological Factors

Exploration Risk

A play consists of a number of pools and/or prospects that may or may not contain hydrocarbons. Therefore, associated with each prospect is

Table 5.1. Types of Evaluation Methods

Type of data available	*Evaluation method*
No data	Comparative geology
Types of source rock (conceptual plays)	Types of expected products (oil, gas, or heavy oil)
Stratigraphic columnar section (conceptual or real)	Burial history, oil or gas window; timing of generation
Areal extent and volume of source rock	Petroleum system, material balance

an exploration risk that measures the probability of a prospect being a pool. Estimating exploration risk in petroleum resource evaluation is important. Methods for quantifying exploration risks are described later.

Geological factors that determine the accumulation of hydrocarbons include the presence of closure and of reservoir facies, as well as adequate seal, porosity, timing, source, migration, preservation, and recovery. For a specific play, only a few of these factors are recognized as critical to the amount of final accumulation. Consequently, if a prospect located within a sandstone play, for example, were tested, it might prove unsuccessful for any of the following reasons: lack of closure, unfavorable reservoir facies, lack of adequate source or migration path, and/or absence of cap rock.

The frequency of occurrence of a geological factor can be measured from marginal probabilities. For example, if the marginal probability for the presence-of-closure factor is 0.9, there is a 90% chance that prospects drilled will have adequate closure. For a prospect to be a pool, the simultaneous presence of all the geological factors in the prospect is necessary. This requirement leads us to exploration risk analysis.

Methods for Estimating Marginal Probability

When we assess a conceptual play, we begin by formulating a play definition. At this stage, a number of questions emerge:

- Does the play in question exist?
- Does the play have an adequate source?
- Can we recover oil or gas from a play that lies under deep water?
- Is the timing of hydrocarbon generation adequate for the play?

Some geological factors such as source, maturation, and migration, for example, would normally be present throughout a play, but at an early stage of exploration we cannot determine whether these factors are in place. PETRIMES provides ways of handling this type of uncertainty. However, we first need to explore the concepts of play-level and prospect-level geological factors.

Play-Level Geological Factor

The play-level geological factor measures the chance that a geological factor is common to all prospects within a play, and is a regional phenomenon across an entire play. The occurrence of a play-level geological factor is denoted by G (global); the marginal probability of this event is represented by θ_g. White (1980) referred to the play-level geological factor as a *play chance* or *group risk* (Gehman et al., 1981; White and Gehman, 1979).

If a play contains hydrocarbons, then all geological factors are present. Let these factors or events be denoted by $G_1, G_2, \ldots, G_j$. The probability of a play having hydrocarbons is then

$$\begin{aligned} \theta_{gi} &= \mathrm{P}[G_i] \\ &= \mathrm{P}[\text{a play has factor } G_i] \\ &= \mathrm{P}[\text{a geological factor } G_i \text{ is satisfied for all prospects within} \\ &\qquad \text{the play, } i = 1, \ldots, j] \end{aligned} \tag{5.1}$$

For example, G_1 = [adequate source], G_2 = [adequate preservation], ...

If all play-level geological factors exist, then

$$\begin{aligned} \theta_g &= \mathrm{P}[G_1 \cap G_2 \cap \cdots \cap G_j] \\ &= \mathrm{P}[\text{play possessing all factors}] \end{aligned} \tag{5.2}$$

If any of these G_i values do not occur, then the play does not contain hydrocarbons. If $G_1, G_2, \ldots, G_j$ are statistically independent, then the probability of having all play-level geological factors simultaneously is defined as follows:

$$\theta_g = \prod_i^j \theta_{gi} \tag{5.3}$$

This play-level geological factor can be considered a parameter to be estimated from data, or an expression of geological judgment.

White (1980) described a facies-cycle wedge (Fig. 5.1) as a body of sedimentary rock bound above and below either by a regional unconformity or by the top of a major nonmarine tongue. The ideal wedge represents a transgressive–regressive cycle of deposition, including, from base to top, a vertical succession that varies from nonmarine to coarse-textured marine, to fine-textured marine, to coarse-textured marine, and back to nonmarine facies. Exploration plays located

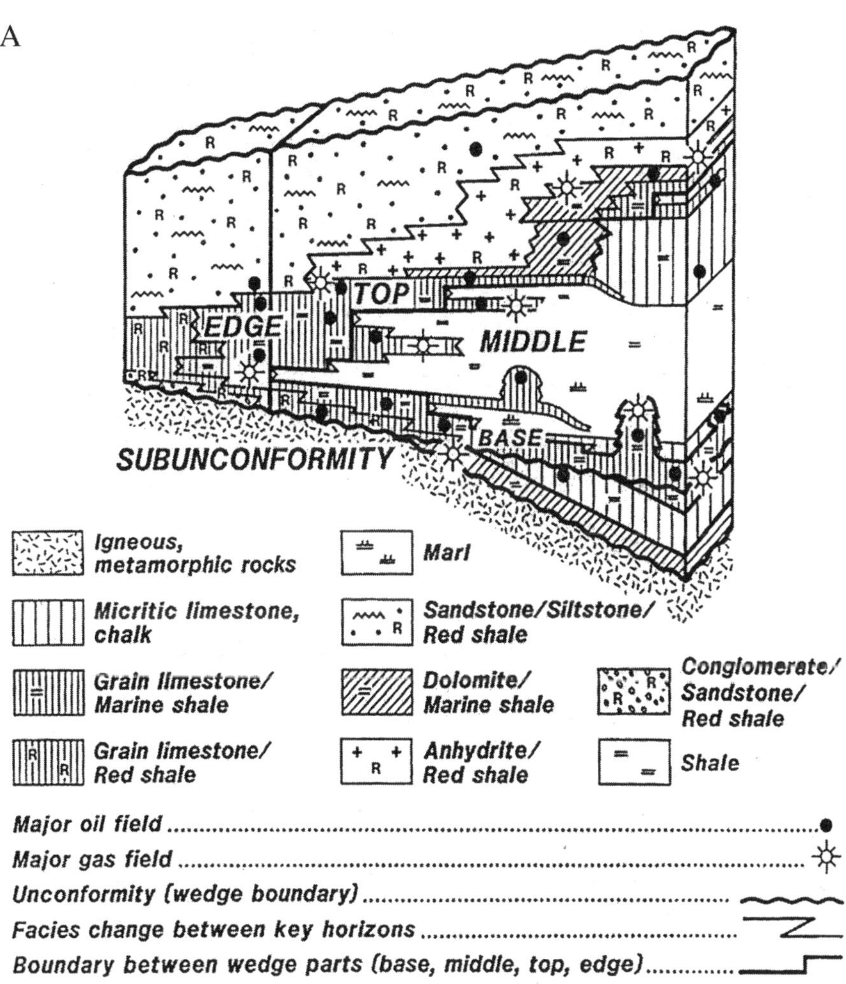

Figure 5.1. (*Continues*)

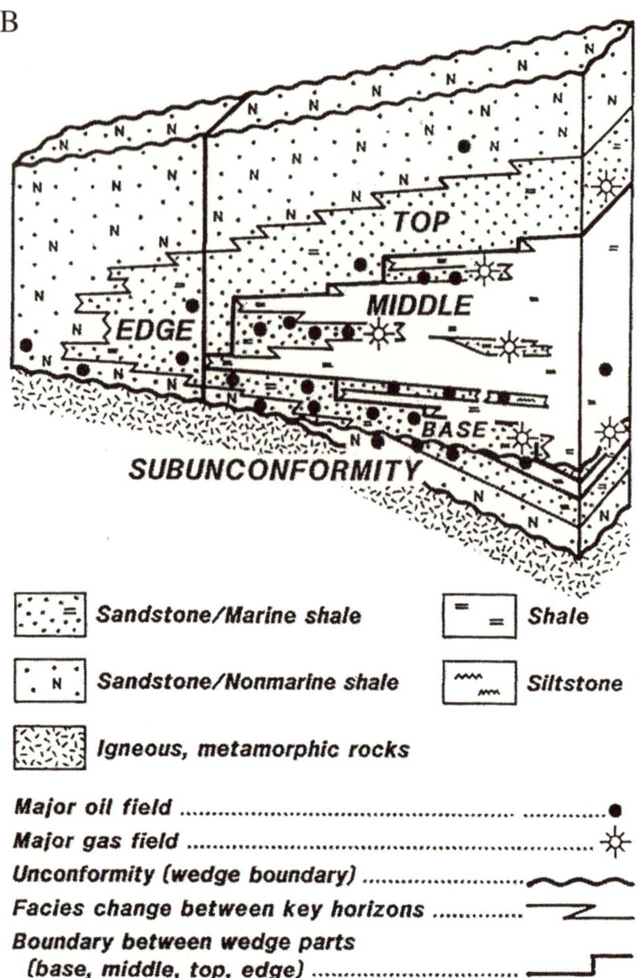

Figure 5.1. (A, B) Types of facies-cycle wedge (after White, 1980). (A) Carbonate–shale wedge. (B) Sand–shale wedge.

within a facies wedge can be allocated into either a wedge-base, wedge-middle, wedge-top, wedge-edge, or subunconformity play. Each such play type is associated with a play-level geological factor. White (1980) summarizes 1150 plays in 80 productive basins of the free world and presents the relationships between play characteristics and the chances of the play containing hydrocarbons. The results are reproduced in Table 5.2.

Table 5.2. Examples of Play-Level Geological Factors for Various Geological Models

Play type	*Example*	*Exploration risk*	
		Sandstone	*Carbonate*
Edge	Eocene to Miocene Cook Inlet, Alaska; Jean Marie*	0.15	0.15
Top	Belly River, Mission Canyon	0.15	0.44
Base	Mannville, Beaverhill Lake	0.60	0.35
Subunconformity	Jurassic, Mississippian	0.45	0.30

*Added by author. After White (1980).

Prospect-Level Geological Factor

The prospect-level geological factor measures the marginal probability that a geological factor exists for an individual prospect. A prospect-level geological factor is represented by R (local), and its marginal probability is denoted by θ_r. The risk can also be considered as a super-population parameter, and can be estimated from data. For the prospect-level geological factor, absence of such factors as closure, reservoir facies, or porosity will result in a prospect lacking hydrocarbons. This, however, does not imply that these factors are also absent from other prospects in the play.

Let $R_1, R_2, \ldots, R_k$ denote the geological factors for an individual prospect at the prospect level. For example:

$$R_1 = [\text{presence of closure}]$$
$$R_2 = [\text{adequate seal}], \ldots, \text{and so on.}$$

Let us define

$$G = G_1 \cap G_2 \cap \ldots G_j$$
$$R = R_1 \cap R_2 \cap \ldots R_k$$

A prospect within a play contains hydrocarbons if, and only if, (1) the play has all play-level geological factor factors and (2) the prospect meets all prospect-level geological factor requirements. In other words, a prospect contains hydrocarbons if, and only if, $G \cap R$.

If we define $\theta_{ri} = \mathrm{P}[R_i|G]$ and $\theta_r = \mathrm{P}[R|G]$, then the probability of hydrocarbons being present is defined as

$$\begin{aligned}&\mathrm{P}[\text{a prospect containing hydrocarbon}]\\&\quad = \mathrm{P}[G \cap R] = \mathrm{P}[R|G] \times \mathrm{P}[G]\\&\quad = \theta_r \times \theta_g\end{aligned} \tag{5.4}$$

If the geological factors are independent, then the prospect-level geological factor is defined as

$$\theta_r = \prod_i^k \theta_{ri} \tag{5.5}$$

If the risk factors are not independent, then the rule of multiplication of the conditional probability rule must be applied as follows:

$$\theta_r = \mathrm{P}[R_1 \cap R_2 \cap \cdots \cap R_k] \tag{5.6}$$

Integrating information obtained from tested wells with data from adjacent wells can identify the presence or absence of a particular prospect-level geological factor. For example, the presence or absence of closure can be recognized by reviewing stratigraphic or seismic correlations after drilling. The existence of reservoir facies can be identified from mechanical logs. Adequacy of seal can be established by examining (1) the presence or absence of cap rock, (2) the quality of the seal, and (3) possible leakage of the closure. Adequate source and migration factors mean that oil has migrated into the trap. Therefore, if a potential reservoir is shown from drill stem tests to contain either oil, oil shows, or oil traces, then the factor is considered to be present.

Marginal Probability Distribution

Figure 5.2A displays a probability distribution for the geological factor of adequate maturation. The assumption used here is that either the sample size is large enough to represent the play (population), or it is a random sample from the play (population). We also assume that the geochemical interpretations are valid.

The distribution suggests a 70% chance that the percentage of hydrocarbons extracted from the play in question would range from

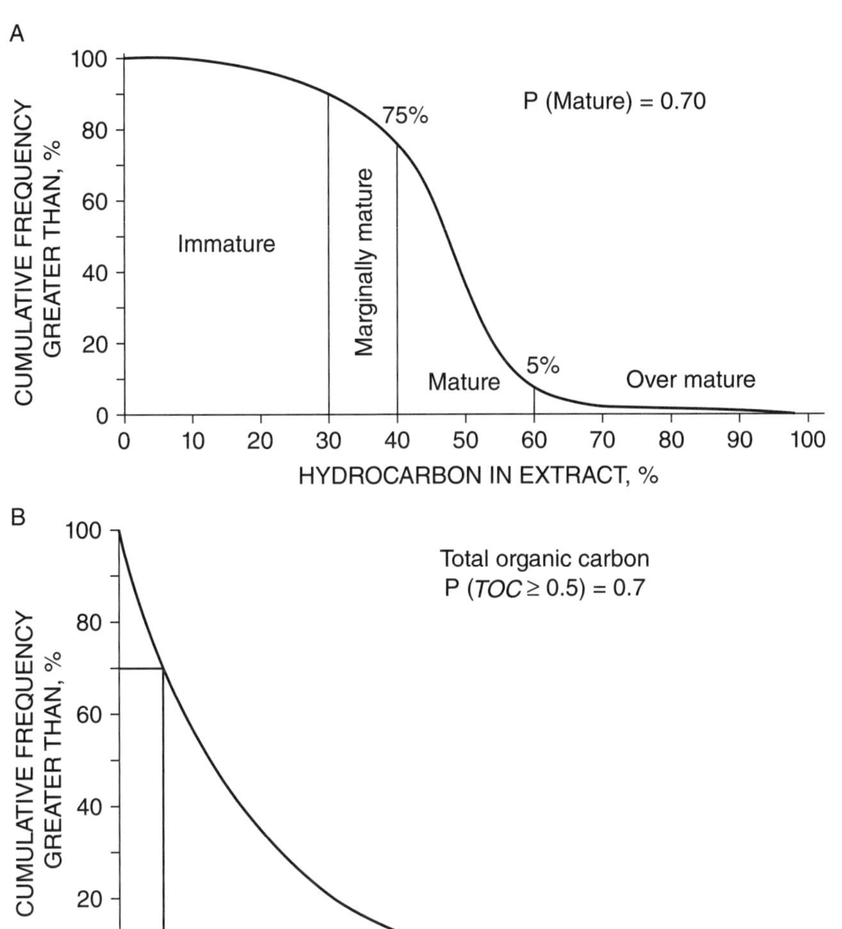

Figure 5.2. (A, B) Example probability distributions for a random variable of source rock maturation (A) and amount of total organic matter (B). *TOC*, total organic carbon.

40% to 60%. This would reflect a mature source rock and be defined as follows:

$$P[40\% \leq \text{mature} \leq 60\%] = 0.70$$

Figure 5.2B displays the probability distribution for total organic carbon. From this distribution, there is a 70% chance that the play has a total organic carbon content in excess of 0.5. The marginal probability for adequate source is interpreted as 0.7—in other words,

$$\mathrm{P}[TOC \geq 0.5\%] = 0.7$$

where TOC is total organic carbon.

Dependence in Prospect-Level Geological Factors

Traditionally, exploration risk is an expression of the products of marginal probabilities of geological factors, such as the presence of closure, reservoir facies, adequate source, and adequate seal. The statistical assumption presumed in such a product operation is that risk factors are independent. The assumption of independence of risk factors has been challenged using exploratory well data obtained from the Huang-Hua Basin of eastern China.

Data from 242 exploratory wells in a sandstone play in the Huang-Hua Basin were analyzed to determine why a particular well had failed. The presence or absence of closure and of reservoir facies, as well as the adequacy of source and seal, were recorded for each well (Lee et al., 1989). In Table 5.3, the number one indicates that a factor is present, whereas zero indicates that a factor is absent.

First, if we assume that these factors are independent of each other, then the overall prospect-level geological factor is the product of 184/242, 220/242, 185/242, and 228/242, which equals 0.50.

Second, the geological factors were analyzed using the following conditional probability formula:

$$\begin{aligned}
&\mathrm{P}[\text{Closure} \cap \text{Reservoir Facies} \cap \text{Source} \cap \text{Seal}] \\
&\quad = \mathrm{P}[\text{Closure}] \times \\
&\qquad \mathrm{P}[\text{Reservoir Facies} \mid \text{Closure}] \times \\
&\qquad \mathrm{P}[\text{Source} \mid \text{Closure} \cap \text{Reservoir Facies}] \times \\
&\qquad \mathrm{P}[\text{Seal} \mid \text{Closure} \cap \text{Reservoir Facies} \cap \text{Source}] \\
&\quad = 184/242 \times 127/184 \times 111/127 \times 109/111 = 0.45
\end{aligned} \tag{5.7}$$

The difference between these two approaches is 0.05. This example demonstrates that geological factors might not be independent. The dependency between any two factors was studied further by using

Table 5.3. Example of Data Set for Exploration Risk Analysis

Closure	*Reservoir facies*	*Migration*	*Source*
1	0	1	1
0	1	1	1
1	1	0	1
1	1	1	1

0, absent; 1, present.

chi-square tests that indicated three pairs of factors (closure and source, closure and seal, and facies and source) were dependent factors, whereas other pairs were independent. The data set was also subjected to correlation analysis. For all dependent pairs of factors, significant correlation was established.

The East Coast Play

Table 5.4 displays the factors and their marginal probabilities for a conceptual play from the East Coast of Canada. The assessor interpreted the geological factors as either play-level or prospect-level geological factors. The first column displays the names of the geological factors; the second column shows the corresponding marginal probability. The

Table 5.4. Marginal Probabilities Used to Calculate Exploration Risk for the East Coast Conceptual Play

Geological factor	*Marginal probability*	*Case*	
		I	*II*
Presence of closure	0.95	Prospect	Prospect
Presence of facies	0.90	Prospect	Prospect
Adequate timing	0.95	Play	Play
Adequate seal	0.80	Prospect	Prospect
Adequate source	0.75	Prospect	Play
Adequate preservation	0.80	Prospect	Play
Overall play-level geological factor		0.95	0.57
Overall prospect-level geological factor		0.41	0.68
Exploration risk		0.39	0.39

last two columns display the interpretations of each factor as prospect level or play level. For case I, only the adequate timing factor is considered as a play-level geological factor, whereas in case II, adequate timing, adequate source, and adequate preservation factors are considered as play-level geological factors. There is no information to suggest whether these factors are dependent. Therefore, the overall play-level geological factor is calculated from the multiplication of all play-level marginal probabilities, whereas the overall prospect-level geological factor is the product of all prospect-level marginal probabilities. Finally, the exploration risk is the product of overall play and prospect levels. As seen in Table 5.4, the two overall risks are very different for these two cases. However, the exploration risk is identical. Because of the difference in play-level and prospect-level geological factors, subsequent estimations will vary accordingly.

Pool-Size Distribution

In reservoir engineering, a pool size can be calculated by using the following equation:

$$\begin{aligned}\text{Pool size} = {} & \text{Constant} \times \text{Pool Area} \times \text{Net Pay} \times \\ & \text{Porosity} \times \text{Hydrocarbon Saturation} \times \\ & \text{Recovery Factor/Gas or Oil Formation Volume Factor}\end{aligned} \tag{5.8}$$

For resource evaluation, Equation 5.8 is adapted to define pool-size distribution (Roy, 1979). To solve the equation, the various distributions are multiplied together. This type of multiplication can be accomplished using the Monte Carlo method or an operation of lognormal distributions that approximate the geological random variables.

The Monte Carlo Method

In the 1950s, a procedure known as the Monte Carlo method was used to solve certain types of mathematical problems. Here, Figure 5.3 displays three examples that illustrate how various numerical procedures can be applied to different problems. The first example calculates the area under the line, $Y = X$ (Fig. 5.3A). One can consider that the triangle is located within a square with a unit length. The area beneath the straight line equals half the unit. On the other hand, the area can also

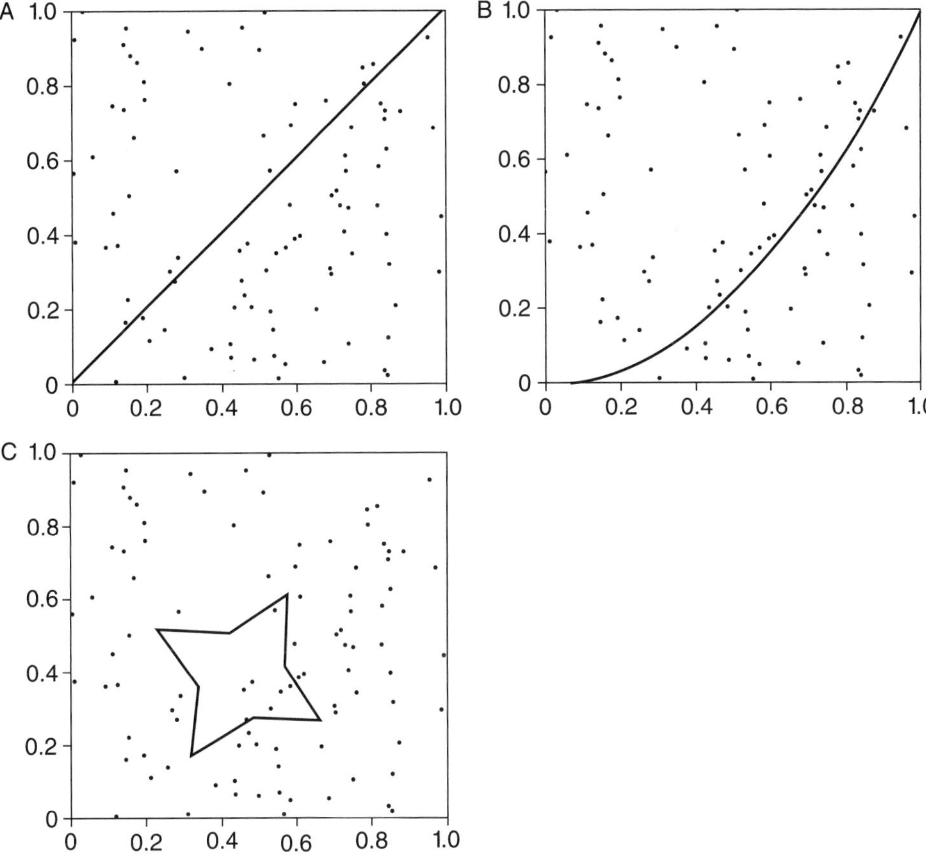

Figure 5.3. (A–C) Examples illustrating the use of the Monte Carlo simulation procedure to compute (A) a lower triangular area; (B) an area under the curve, $y = x^2$; and (C) an area within the polygon.

be estimated using the Monte Carlo method. The steps involved are as follows:

1. Generate a pair of independent random numbers.
2. Use the pair of random numbers as coordinates of a point located within the square.
3. Repeat the procedure N times and plot all points on the square.
4. Count the number of points located within the triangle (i.e., n points).
5. Compute the area of the triangle n/N, which is $59/100 = 0.590$ units in this case.

For this simple example, the Monte Carlo method proves cumbersome.

In the second example (Fig. 5.3B), the area under the $Y = X^2$ curve is calculated from the integration of the curve as follows:

$$Y = \int_0^1 X^2 dx = 1/3$$

Here, the Monte Carlo method can be applied N times where n points are located under the curve. Therefore, the area will be $34/100 = 0.340$ units. The integration method is more efficient than the Monte Carlo method.

The third example (Fig. 5.3C) is used to calculate the polygonal area, which can be calculated by Green's theorem. In this particular case, the Monte Carlo method (the area = 7/100 units, the actual area =1/12 units) is the most efficient.

We can assess the accuracy of the Monte Carlo method by increasing the number of random numbers to 1000, and find that the three areas are equal to 0.509, 0.329, and 0.076 units respectively. It can be observed that the accuracy for each example increases but varies. This is why the Monte Carlo method requires a large sample size to reduce the measurement error.

Atwater (1956) calculated success ratios and average pool sizes from onshore Louisiana, and then estimated the number of prospects in the adjacent offshore. He claimed that the petroleum resources of offshore Louisiana could be approximated from the product of the success ratio, the average pool size, and the number of prospects. The assumptions for this approach are that the average pool size and the success ratio are the same for both offshore and onshore Louisiana. This approach was the basis for the logic of the petroleum resource assessment procedure using the Monte Carlo method.

In the late 1960s, the petroleum industry began to use the computer as a tool for evaluating hydrocarbon plays. For many years, the Monte Carlo procedure has been used in play estimation (Energy, Mines and Resources, 1977; White and Gehman, 1979), and has been widely used in petroleum resource evaluation articles since then.

Figure 5.4 illustrates how to use the Monte Carlo method to compute pool-size distribution. Geological variables (right side of the equation) are expressed by their own probability distributions. Random numbers were independently generated as $R_1, R_2, \ldots, R_5$, because there was no information on relationships between variables. These five random

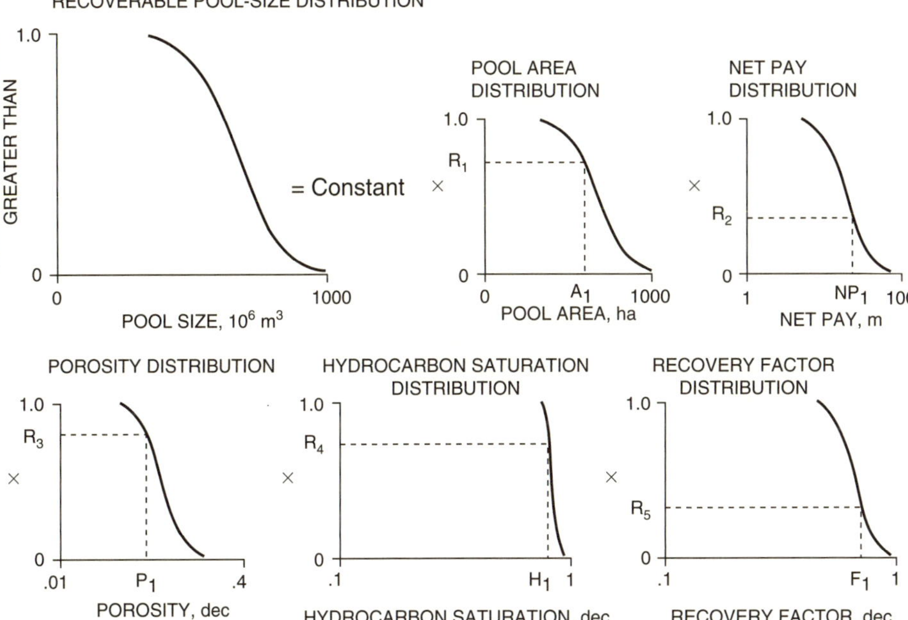

Figure 5.4. Diagram illustrating the Monte Carlo procedure for computing pool-size distribution.

numbers were then multiplied as PS_1. If one repeats this step many times, then all the PS's can be used to construct a pool-size distribution (Fig. 5.4, left side of the equation).

The Lognormal Approximation

Lognormal approximation also can be applied to solve Equation 5.8. In PETRIMES, the geological random variables are jointly approximated through the use of a multivariate lognormal distribution. Because the result of the product and/or division of lognormal random variables is again a lognormal variable (Aitchison and Brown, 1973), it follows that the pool-size distribution is lognormal. If we let μ_i, σ_i^2, and σ_{ij}, i, $j = 1, 2, \ldots,$ denote the mean, variance, and covariance of the natural logarithms of the geological variables, then the mean and variance of the pool-size distribution are given by

$$\text{Mean} = e^{(\mu + \sigma^2/2)} \tag{5.9}$$

$$\text{Variance} = e^{(2\mu + \sigma^2)} \times (e^{\sigma^2} - 1) \tag{5.10}$$

$$\mu = \ln(\text{Constant}) + \sum_i \mu_i \tag{5.11}$$

$$\sigma^2 = \sum \sigma_i^2 + 2\sum_{i<j}\sum \sigma_{ij} \tag{5.12}$$

Equation 5.8 can either be applied to mature, immature, or conceptual plays. For conceptual plays, we have no discovery record to apply to the discovery process model. The pool-size equation can then be used to derive pool-size distribution, as shown in Equation 5.8. Furthermore, distributions of variables such as pool area and net pay are based on interpretations by geologists and/or on comparative studies. These are considered to be superpopulation distributions.

Examples

The Beaverhill Lake Play

The Beaverhill Lake play is used here to demonstrate the application of the pool-size equation approach (Eq. 5.8) when a large number of discoveries are available. For this play, variations in hydrocarbon saturation and the oil shrinkage factor are relatively small compared with other variables. Also, no significant correlation exists between hydrocarbon saturation and the oil shrinkage factor and other variables. Therefore, they are not included in the total variance. If we then only consider pool area, average net pay, and average porosity, Equation 5.8 is reduced to

$$\begin{aligned}&\text{Oil pool size in place } (10^6\,\text{m}^3)\\&\quad = \text{Constant} \times \text{Pool Area} \times \text{Net Pay} \times \text{Porosity}\end{aligned} \tag{5.13}$$

where the constant equals 0.00681, which is the product of average hydrocarbon saturation, average oil shrinkage factor, and the conversion factor from hectare-meter to million cubic meters.

The reason for computing the oil-in-place is that enhanced oil recovery techniques have been applied to some, but not all, of the pools. Thus, the recovery factor for the play varies from a few percent to as much as 25%. Incorporation of the recovery factor here will introduce an inconsistent measurement of pool size. Nevertheless, PETRIMES will be able to handle all variables in Equation 5.8.

Detailed information for each geological random variable is given in Table 5.5. (Raw data were obtained from the report by the Energy

Table 5.5. Lognormal Parameters and Correlations of Geological Variables for the Beaverhill Lake Play

Variable	*Sample mean* $\hat{\mu}$	*Variance* $\hat{\sigma}^2$	*Correlation**		
			Pool area	*Average net pay*	*Average porosity*
Pool area	7.869	0.721	1.000		
Average net pay	2.211	0.422	0.682 (0.731)	1.000	
Average porosity	–2.674	0.068	0.641 (0.275)	0.452 (0.077)	1.000

*Covariance in parentheses.

Constant = 0.681 $2\sum_i\sum_j \sigma_{ij} = 2.164,\ i < j$

Scale factor = 0.001

$\hat{\mu} = 2.408,\ \hat{\sigma}^2 = 3.211 + 2.164 = 5.375$

Resources Conservation Board [1989, Table 2–5].) From Table 5.5 we can see that because the pool area contributes most to the values of μ and σ^2, it is the most important random variable contributing to the pool-size equation. Correlation and covariance for the three random variables are also given in Table 5.5. The pool area and average net pay random variables (Fig. 2.11B), as well as porosity and pool area, have high correlation coefficients of 0.682 and 0.641 respectively. In this example, if the covariances are incorporated, the mean of the pool size will be 151×10^6 m^3 of oil. In contrast, if they are all ignored, the mean is reduced to 46×10^6 m^3 of oil. Similarly, if negative correlations are omitted, then the mean will be overestimated.

The advantages of using Equation 5.8 are that (1) we can gain a better understanding of the variables, their interdependence, and their influence on pool-size distribution; and (2) geological random variables for an undiscovered pool, such as pool area and average net pay, can also be regenerated for a given pool size (see "Generation of Reservoir Parameters" later in this chapter).

Moreover, because we usually do not have sufficient data to compute covariances of geological random variables for conceptual plays, the variance of pool-size distribution can be under- or overestimated. Furthermore, correlations of random variables can change from population to population. For example, log–log relationships between porosity and water saturation for the Bashaw reef (Fig. 5.5A) and Cardium marine sandstone (Fig. 5.5B) display distinct correlation patterns. Examining possible correlations might lead to justifying the

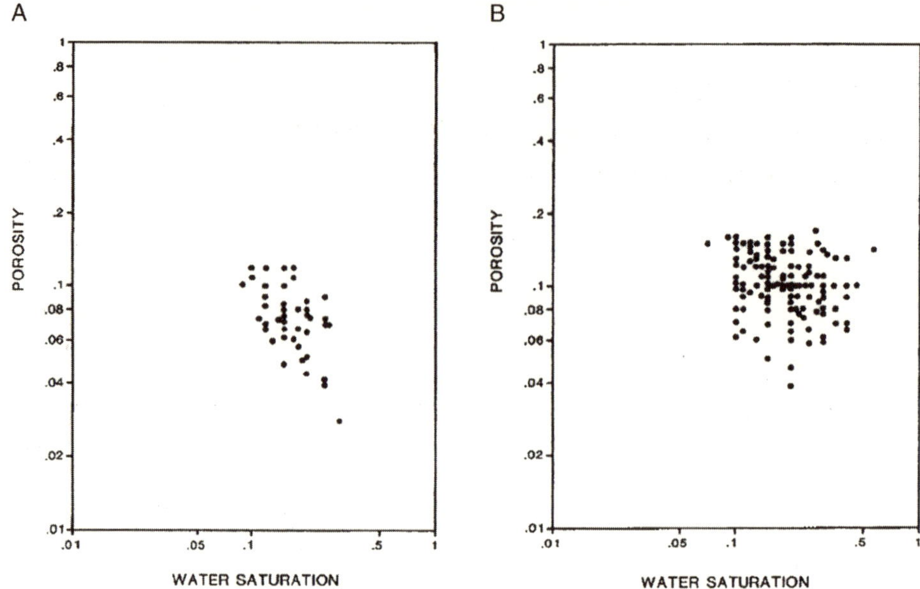

Figure 5.5. (A, B) Log-linear relationships between water saturation and porosity of Bashaw play (A) and Cardium play (B), Western Canada Sedimentary Basin.

addition or subtraction of the variance from the pool-size distribution. If a lognormal distribution were adopted, the variance and covariance could be adjusted. The Beaverhill Lake play was used to describe the roles of Equations 5.11 and 5.12 in the pool-size equation.

Figure 5.6A displays the positive correlation between the random variables of pool area and average net pay. Figure 5.6B demonstrates the impact of the covariance to the pool-size distribution. The solid line of Figure 5.6B shows the pool-size distribution derived by omitting the covariance, whereas the circles show the pool-size distribution derived by including the covariance. The former distribution has a mean of 13×10^6 m^3 and the latter has a mean of 32×10^6 m^3. The difference is more than double.

Figure 5.7 demonstrates the impact of negative covariance on the pool size. The solid line shows the pool-size distribution derived by omitting the negative covariance between the pool area and average net pay, whereas the circles show the pool-size distribution derived by including the covariance. The difference of the two means is more than double.

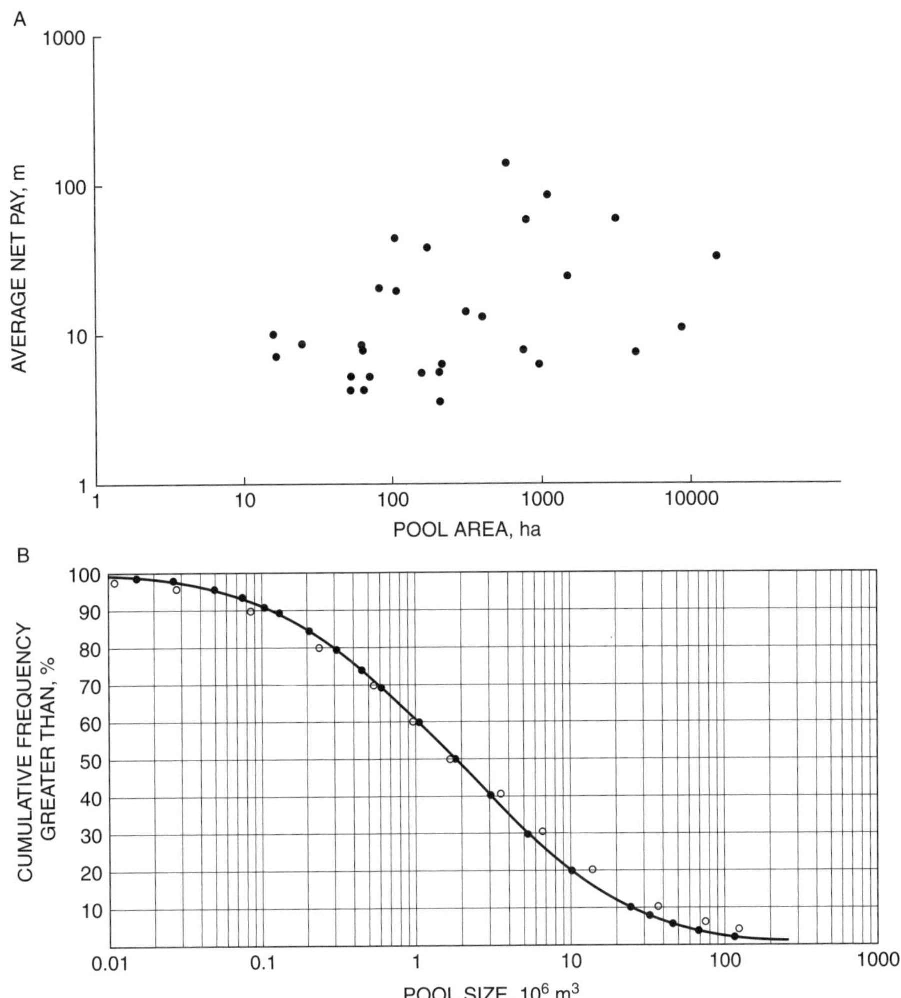

Figure 5.6. (A, B) Diagrams showing the correlation between random variables of pool area in hectares and average net pay in meters (A), and the pool-size distribution (B). The solid line indicates the pool-size distribution (mean = 13×10^6 m^3) derived by omitting the covariance shown in (A). Circles indicate the pool-size distribution (mean = 32×10^6 m^3) derived by including the covariance between pool area and average net pay. Data from the Western Canada Sedimentary Basin.

The preceding examples demonstrated the impact of correlation on the mean of a pool-size distribution when the sample covariance matrix had been computed and used. The population covariance matrix should be computed using MDSCV (see Chapter 3).

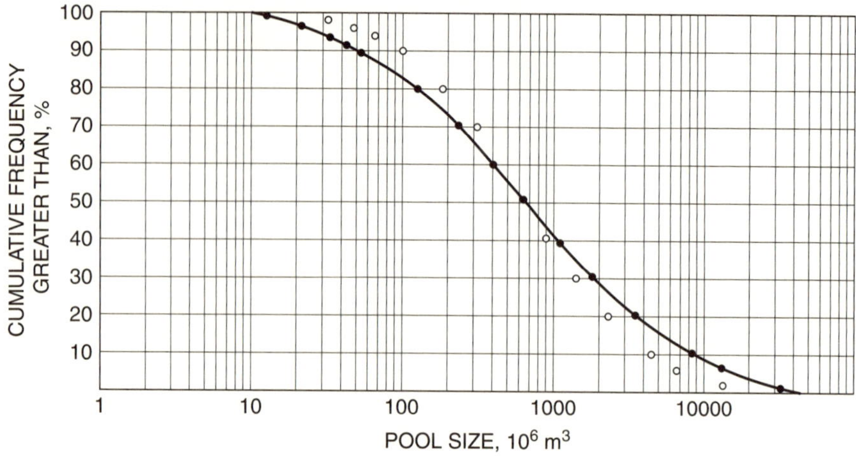

Figure 5.7. Pool-size distributions of the Bashaw reef play. The solid line is derived by omitting the negative covariance between the average net pay in meters and the pool area in hectares (mean = 4497×10^6 m^3). The circles were derived by including the negative covariance between the average net pay in meters and the pool area in hectares (mean = 2069×10^6 m^3). Data from the Western Canada Sedimentary Basin

The East Coast Play

One play from the East Coast of Canada was selected to illustrate the application of PETRIMES in a conceptual play. The data used include probability distributions of area of closure, reservoir thickness, porosity, and trap fill. The equation used to calculate pool-size distribution is as follows:

$$\text{Pool Size} = c \times \text{Area of Closure} \times \text{Reservoir Thickness} \times \text{Porosity} \times \text{Trap Fill} \tag{5.14}$$

where c is the product of hydrocarbon saturation and a conversion factor of cubic feet to millions of barrels. Pool size is oil-in-place measured in MMbbls.

The probability distributions for reservoir thickness, porosity, and trap fill are considered as superpopulation distributions. The distribution of the area of closure, for example, was derived from structural contour maps based on seismic data. The distribution proposed by geologists was plotted as a solid line in Figure 5.8, and was approximated by a lognormal distribution (indicated by open circles).

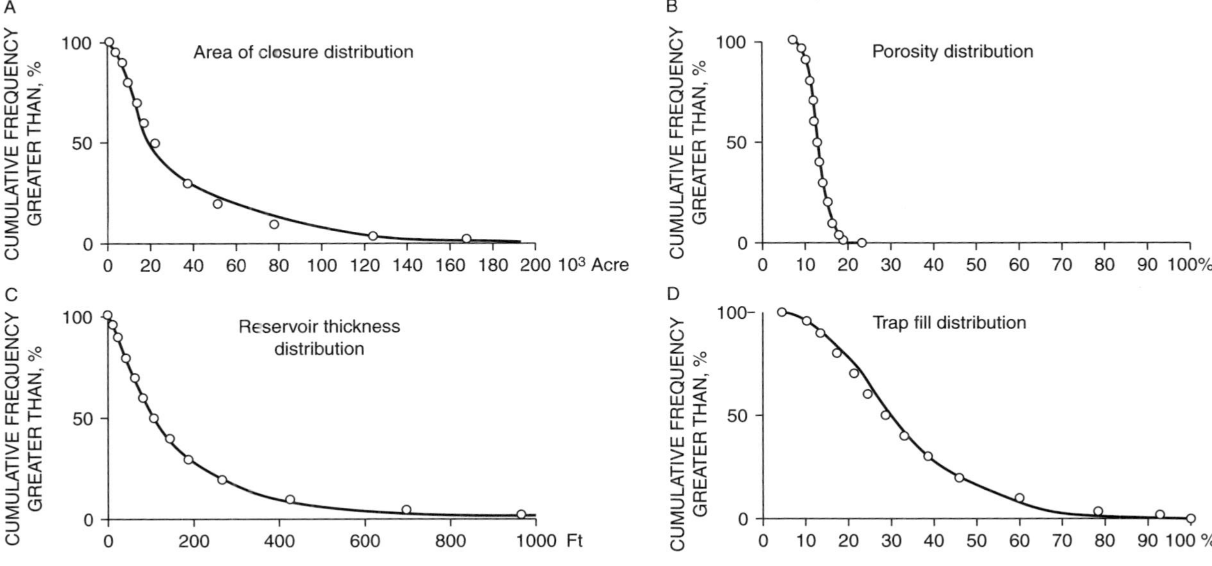

Figure 5.8. (A–D) Distributions of area of closure (A), reservoir thickness (C), porosity (B), and trap fill (D) for the East Coast play, Canada.

If the geological variables are approximated by lognormal distributions with parameters μ and σ^2, and if they are independent, then

$$\ln x = \ln c + \sum \ln Z_i \tag{5.15}$$

is normally distributed with $\hat{\mu} = 2.882$ and $\hat{\sigma}^2 = 2.5$, and its density is given by

$$h(x) = \frac{1}{x\sigma\sqrt{2\pi}} \exp\left[-\frac{1}{2}\left(\frac{\ln x - \mu}{\sigma}\right)^2\right] \tag{5.16}$$

where x is the pool size in MMbbls.

Values calculated by Equation 5.16 were plotted as circles in Figure 5.9. The pool-size distribution, plotted as a solid line in Figure 5.9, was derived using the Monte Carlo approach based on the original four distributions.

In this example, the pool-size distribution derived from the Monte Carlo simulation resembles the lognormal distribution, except at the 0.5% level. The Monte Carlo simulation usually yields a less skewed distribution, whereas a lognormal approximation extends the tail of the distribution.

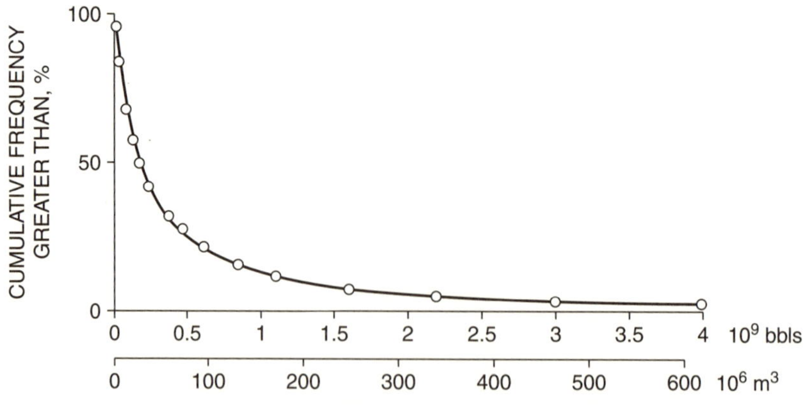

Figure 5.9. Pool-size distribution of the East Coast play. The circles indicate the distribution derived by lognormal approximation. The solid line indicates the distribution derived by the Monte Carlo procedure (input distributions are displayed in Fig. 5.8).

Estimating Resources

Number-of-Prospects Distribution

If an identifiable type of trap, such as an anticline, can be mapped on the surface of a play or detected seismically at depth, then the number of prospects can be counted. Some of the prospects cannot be mapped on the surface because of the presence of vegetation. They also might not be detected at depth because seismic coverage might be too sparse to detect small prospects. Three questions might arise at this point:

1. What is the maximum number of prospects that the play could have?
2. Given a 50% chance, what is the least number of prospects that the play could have?
3. What is the observed number of prospects?

From the answers to these questions, one can construct a number-of-prospects distribution that can be considered a superpopulation distribution. Figure 5.10 displays an example of a number-of-prospects distribution for a conceptual play. The mean and variance of the distribution are 103 and 77.09 respectively. Given a 50% chance, the play will have more than 100 prospects.

Number-of-Pools Distribution

The number-of-prospects distribution will be used with exploration risk to derive the number-of-pools distribution. Let M be the random

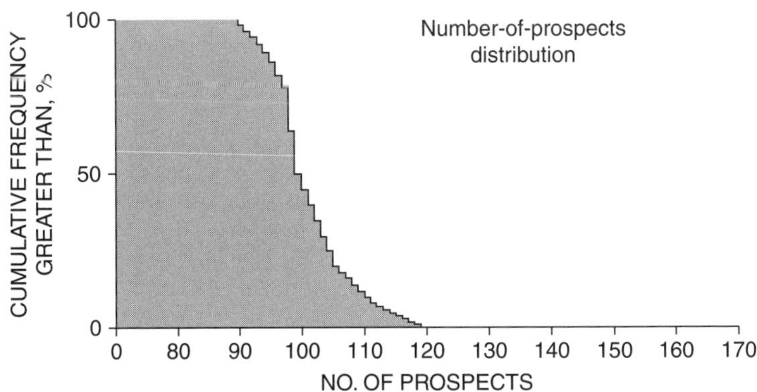

Figure 5.10. Number-of-prospects distribution for the East Coast play, Canada (after Lee and Wang, 1983b).

variable denoting the total number of prospects in a play and m be a value of M. Let its probability function be

$$\mathrm{P}[m] = \mathrm{P}[M = m], \quad m = m_0, \ldots, m_i$$

This distribution could be obtained from seismic detection and expert knowledge of the play. Associated with the ith prospect, we define

$$I = \begin{cases} 1, & \text{if the } i\text{th prospect satisfies the condition } R \\ 0, & \text{otherwise} \end{cases}$$

Given that event G has occurred (i.e., the play has all the conditions necessary for hydrocarbon occurrence), the total number of pools in the play is given as

$$N = I_1 + I_2 + \cdots + I_m$$

N is a sum of random variables; therefore, its conditional probability distribution, given G, is

$$\begin{aligned} &\mathrm{P}\left[N = n|G\right] \\ &\quad = \sum_m \mathrm{P}\left[N = n, M = m|G\right] \\ &\quad = \sum_m \mathrm{P}\left[N = n|M = m, G\right] \times \mathrm{P}\left[M = m\right] \\ &\quad = \sum_m \mathrm{P}\left[I_1 + I_2 + \cdots + I_m = n|M = m, G\right] \times \mathrm{P}\left[m\right] \end{aligned} \tag{5.17}$$

where N is the random variable for the number of pools and n is a specific value for N. We have assumed $[M = m]$ is statistically independent of G for all m. Moreover, we assume $I_1, I_2, \ldots$ are independent of M and all I_i's are also independent.

Because $\mathrm{P}[I_i = 1|\,G] = q_r$ for all i, then

$$\mathrm{P}\left[N = n|\,G\,\right] = \sum_m \binom{m}{n} q_r^n \left(1 - q_r\right)^{m-n} \mathrm{P}[m], \quad \text{for } n = 0, \ldots, m_1$$

The sum extends from $m = \max(n, m_0)$ to m_1. Denote as G^T the complement of G. The distribution of N is now given as

$$\mathrm{P}[N=n]=\mathrm{P}\left[N=n\,|\,G\right]\times\mathrm{P}[G]+\mathrm{P}\left[N=n\,|\,G^{T}\right]\times\mathrm{P}\left[G^{T}\right]$$
$$=\begin{cases}(1-\theta_g)+\theta_g\sum\limits_m(1-\theta_r)^m\,\mathrm{P}[m], & \text{if } n=0\\ \theta_g\sum\limits_m\binom{m}{n}\theta_r^n\,(1-\theta_r)^{m-n}\,\mathrm{P}[m], & \text{if } n\geq 1\end{cases} \tag{5.18}$$

Also,

$$\begin{aligned}\mathrm{P}[\text{play has at least one pool}]&=\mathrm{P}[N\geq 1]\\ &=1-\mathrm{P}[N=0]\\ &=\theta_g\left[1-\sum_m(1-\theta_r)^m\,\mathrm{P}[m]\right]\end{aligned} \tag{5.19}$$

For example, case II gives $\theta_g = 0.57$ and $\theta_r = 0.68$. If $M = 6$, then

$$\begin{aligned}\mathrm{P}[N\geq 1]&=0.57\left[1-(1-0.68)^6\right]=0.57, \text{ and}\\ \mathrm{P}[N=0]&=1-0.57=0.43, \text{ or } 43\%\end{aligned}$$

The expected value of N is given as

$$\begin{aligned}\mathrm{E}[N]&=\theta_g\,\mathrm{E}\left[N\,|\,G\right]+(1-\theta_g)\,\mathrm{E}\left[N\,|\,G^{T}\right]\\ &=\theta_g\sum_n n\sum\binom{m}{n}\theta_r^n(1-\theta_r)^{m-n}\,\mathrm{P}[m]\\ &=\theta_g\sum_m m\theta_r\sum_n\binom{m-1}{n-1}\theta_r^{n-1}(1-\theta_r)^{m-n}\,\mathrm{P}[m]\end{aligned} \tag{5.20}$$

Therefore,

$$\begin{aligned}\mathrm{E}[N]&=\theta_g\times\theta_r\times\mathrm{E}[M]\\ \mathrm{E}[M]&=\sum_m m\,\mathrm{P}[m]\\ &=\text{expected number of prospects}\end{aligned} \tag{5.21}$$

Similarly,

$$\begin{aligned}
\mathrm{E}\left[N^2 \mid G\right] &= \sum_m \sum_n n^2 \binom{m}{n} \theta_r^n (1-\theta_r)^{m-n}\, \mathrm{P}[m] \\
&= \sum_m \left[m \times \theta_r (1-\theta_r) + m^2 \times \theta_r^2\right] \times \mathrm{P}[m] \\
&= \theta_r \times (1-\theta_r) \times \mathrm{E}[m] + \theta_g^2 \times \sigma_M^2 + \theta_r^2 \times (\mathrm{E}[M])^2
\end{aligned}$$

Hence,

$$\begin{aligned}
\mathrm{Var}[N] &= \mathrm{E}\left[N^2\right] - \left[\mathrm{E}(N)\right]^2 \\
&= \theta_g\, \mathrm{E}\left[N^2 \mid G\right] - \theta_r^2 \times \theta_g^2 \times \mathrm{E}[M]^2 \\
&= \theta_g \times \theta_r^2 \times \mathrm{E}[M]^2 - \theta_g^2 \times \theta_r^2 \times \mathrm{E}[M]^2 \\
&= \theta_g \times \theta_r (1-\theta_r) \times \mathrm{E}[M] + \theta_g \times \theta_r^2 \times \sigma_M^2
\end{aligned} \tag{5.22}$$

Therefore,

$$\sigma_N^2 = \theta_g \times \theta_r \left[\theta_r (1-\theta_g) \times \mathrm{E}[M]^2 + (1-\theta_r) \times \mathrm{E}[M] + \theta_r \times \sigma_M^2\right] \tag{5.23}$$

Equation 5.23 shows that σ_N^2 is dominated by E[M], because the contribution from σ_M^2 is diminished by the multiplier θ_r.

The number-of-prospects distribution (Fig. 5.8) and the risks for case I and case II (Table 5.6) were applied to derive the number-of-pools distribution. From the results (Table 5.7) we can conclude that

Table 5.6. Exploration Risk for the Conceptual Play

Geological factor	*Marginal probability*	*Case*	
		I	*II*
Presence of closure	0.95	Prospect	Prospect
Presence of facies	0.90	Prospect	Prospect
Adequate timing	0.95	Play	Play
Adequate seal	0.80	Prospect	Prospect
Adequate source	0.75	Prospect	Play
Adequate preservation	0.80	Prospect	Play
Overall play-level geological factor		0.95	0.95
Overall prospect-level geological factor		0.41	0.68
Exploration risk		0.39	0.39

Table 5.7. Number-of-Pools Distribution for the Two Cases

Upper percentile	*Number of pools*	
	Case I	*Case II*
0.95	0	0
0.90	33	0
0.75	42	0
0.57	42	0
0.50	42	62
0.25	46	71
0.10	50	78
0.00	80	102

1. their means are identical, but case II has a much larger variance;
2. given a 50% chance, the play will have more than 42 pools for case I and 62 pools for case II;
3. for case I, there is about a 5% chance that the play has no pools, whereas for case II the chance for no pools is about 57%;
4. case II is interpreted as a very risky play

Play Resource Distribution

The operation using a number-of-pools distribution and a pool-size distribution will yield a play resource distribution. The play resource distribution is defined as

$$T = X_1 + X_2 + \cdots + X_N = \sum_i^N X_i \tag{5.24}$$

The play potential distribution is discontinuous at zero, as follows:

$$\begin{aligned} \mathrm{P} = [T = 0] &= \mathrm{P}[N = 0] \\ &= \mathrm{P}[\text{no pools}] \\ &= (1 - \theta_g) + \theta_g \sum_m (1 - \theta_r)^m \, \mathrm{P}[m] \end{aligned} \tag{5.25}$$

Now, for $t > 0$, the greater-than cumulative density function of T is

$$\begin{aligned} F_T(t) &= \mathrm{P}[\text{play resource } > t] \\ &= \mathrm{P}[T > t] \\ &= \sum_{n=1}^{m_1} F_n(t) \, \mathrm{P}[N = n] \end{aligned}$$

where $F_n(t) = \mathrm{P}[X_1 + X_2 + \cdots + X_n > t]$.

The probability function of T is given as

$$f_n(t) = \begin{cases} \mathrm{P}[N=0], & \text{if } t=0 \\ \sum_{n=1}^{m_1} f_n(t)\,\mathrm{P}[N=n], & \text{if } t>0 \end{cases}$$

where $f_n(t)$ is the probability density function of the convolution $X_1 + \cdots + X_n$ of n pool sizes.

The expected value and variance of T are

$$\mathrm{E}[T] = \mathrm{E}[X] \times \mathrm{E}[N] = \theta_g \times \theta_r \times \mathrm{E}[M] \times \mathrm{E}[X] \tag{5.26}$$

$$\sigma_T^2 = \sigma_X^2 \times \mathrm{E}[N] + (\mathrm{E}[X])^2 \times \sigma_N^2 \tag{5.27}$$

where E[N] is the mean of the number-of-pools distribution, E[X] is the mean of the pool-size distribution, σ^2 is the variance of the pool-size distribution, and σ_N^2 is the variance of the number-of-pools distribution.

If X is lognormally distributed with μ and σ^2, then

$$\mathrm{E}[T] = \mathrm{E}[N] \times \exp\left[\left(\frac{\mu + \sigma^2}{2}\right)\right] \tag{5.28}$$

The uncertainty of the play resource distribution as measured by its variance is relatively insensitive to the uncertainty inherited from the prospect distribution. This can be examined by substituting σ_N^2 from Equation 5.23 into Equation 5.27.

$$\sigma_T^2 = e^{(2\mu + \sigma^2)} \times \left(\mathrm{E}[N] \times e^{\sigma^2 - 1} + \sigma_N^2\right) \tag{5.29}$$

The play resource distribution is the superpopulation distribution of the geological model. The uncertainty in the distribution can be reduced if we have pool sizes and their ranks as discussed in Chapter 3. For frontier plays, we are unable to reduce this type of uncertainty because of the lack of information.

Table 5.8. Play Resource Distributions for the Two Cases

Upper percentile	*Play potential, Bbbls*	
	Case I	*Case II*
0.90	1.22	0
0.80	1.57	0
0.70	1.82	0
0.60	2.05	0
0.55	2.29	2.29
0.50	2.28	2.80
0.40	2.55	3.42
0.30	2.86	3.96
0.20	3.30	4.61
0.10	4.05	5.59
Mean	2.50	2.50
Standard deviation	1487	2568

The play resource distributions for cases I and II are given in Table 5.8 and are interpreted as follows:

1. The case I and case II means are identical, but case II has a much larger standard deviation than case I.
2. Case I suggests that there is a chance of about 10% that the play has no potential, whereas for case II the chance is about 45%, as indicated by one minus the probability of the first occurrence of play potential (e.g., 1.0−0.55).
3. In Table 5.8, case II has a higher resource at the tail of the play resource distribution than case I. This is because of the geological factors being interpreted differently, as either play-level or prospect-level geological factors. For case II, if source and preservation factors do exist in one prospect, then they also exist in every prospect. This is why the probability of having more potential (if the potential does exist) is higher in case II than in case I.

Pool-Size-by-Rank

For frontier plays, pool-size-by-rank is normally obtained from operations of pool-size and number-of-pools distributions. Because the number-of-pools distribution is used in estimations of individual pool size, the

Table 5.9. Pool-Size-by-Rank for Case I

Rank	*Probability**	*Mean*	*SD*	*Upper percentile*				
				95	*75*	*50*	*25*	*5*
1	0.95	782	1008	183	329	522	883	2157
2	0.95	349	244	122	199	286	422	782
3	0.95	231	129	93	145	200	280	469
4	0.95	171	85	75	114	154	209	331
5	0.95	136	62	62	93	124	165	253
6	0.95	112	49	53	78	103	136	203
7	0.95	95	40	45	67	87	114	168
8	0.95	81	33	39	58	75	98	142
9	0.95	71	28	35	51	66	85	123
10	0.95	62	24	30	45	58	74	107
11	0.95	55	21	27	40	51	66	94
12	0.95	49	19	24	35	46	59	83
13	0.95	44	17	21	32	41	53	75
14	0.95	39	15	19	28	37	47	67
15	0.95	35	14	17	26	33	43	61
16	0.95	32	12	15	23	30	39	55
17	0.95	29	11	14	21	27	35	50
18	0.95	26	10	12	19	25	32	46
19	0.95	24	10	11	17	23	29	42
20	0.95	21.7	8.9	9.7	15.4	20.5	26.7	38.1
21	0.95	19.8	8.2	8.6	13.9	18.7	24.5	35.0
22	0.95	18.1	7.7	6.7	12.6	17.0	22.3	32.2
23	0.95	16.5	7.1	5.8	11.3	15.4	20.5	29.6
24	0.95	15.0	6.7	5.1	10.2	14.1	18.8	27.3
25	0.95	13.6	6.2	4.4	9.2	12.7	17.2	25.2
26	0.95	12.4	5.9	3.8	8.2	11.6	15.7	23.2
27	0.95	11.3	5.5	3.2	7.3	10.5	14.4	21.4
28	0.95	10.2	5.2	2.7	6.5	9.5	13.1	19.8
29	0.94	9.3	4.9	2.3	5.8	8.6	12.1	18.3
30	0.94	8.4	4.6	1.9	5.1	7.8	11.0	16.9
31	0.93	7.7	4.5	1.6	4.5	7.0	10.1	15.6
32	0.92	6.9	4.0	1.3	4.0	6.3	9.2	14.4
33	0.90	6.3	3.8	1.1	3.5	5.7	8.4	13.3
34	0.88	5.7	3.5	1.0	3.1	5.1	7.7	12.3
35	0.86	5.2	3.3	0.8	2.7	4.6	7.0	11.4
36	0.82	4.7	3.1	0.7	2.4	4.2	6.4	10.6
37	0.78	4.3	2.9	0.6	2.1	3.8	5.9	9.8
38	0.74	3.9	2.7	0.6	1.9	3.4	5.4	9.1
39	0.68	3.6	2.5	0.6	1.7	3.1	5.0	8.5
40	0.63	3.3	2.4	0.5	1.6	2.8	4.6	7.9
41	0.57	3.1	2.2	0.5	1.4	2.6	4.2	7.3
42	0.51	2.9	2.1	0.4	1.4	2.4	3.9	6.9

*Probability of *r* pools.
SD, standard deviation.

Table 5.10. Pool-Size-by-Rank for Case II

Rank	*Probability**	*Mean*	*SD*	*Upper percentile*				
				95	*75*	*50*	*25*	*5*
1	0.57	1030	1219	273	464	713	1170	2738
2	0.57	488	311	191	294	409	585	1043
3	0.57	334	168	150	222	296	401	646
4	0.57	257	113	125	179	233	307	468
5	0.57	209	84	107	150	193	149	366
6	0.57	176	66	93	130	164	209	300
7	0.57	152	54	82	113	142	179	253
8	0.57	133	46	74	101	125	157	218
9	0.57	118	39	66	90	112	139	191
10	0.57	106	34	60	81	101	124	169

*Probability of *r* pools.
SD, standard deviation.

probability of having at least *r* pools is provided. The results of the two cases are given in tables 5.9 and 5.10 and can be interpreted as follows:

1. The probability of having at least one pool, or two pools, and so on, is very different for the two cases. For example, the probability of the existence of at least one pool is 0.95 for case I and 0.57 for case II.
2. The sum of the products (of each individual pool-size mean and its probability of existence) equals the mean of the play resource distribution.
3. The estimated pool sizes for case II are much larger than those of case I. This variability is inherent because of the variances in play resource distributions.

Generation of Reservoir Parameters

For economic analysis of petroleum resources, it is necessary to find the conditional distribution of the geological variables in Equation 5.8 for a given pool size, x. For example, the following question might be asked: Given a pool size equal to 714 MMbbls, what is the distribution of its pool area and net pay?

We assume the vector of the geological variables $\mathbf{Z} = (Z_1, Z_2, \ldots, Z_p)$ associated with the pool-size equation

$$x = z_1 \times z_2 \times \cdots \times z_p \tag{5.30}$$

Table 5.11. Reservoir Parameters Conditional on the Pool Sizes

Pool size, MMbbls	*Reservoir parameter*	*Upper percentile*		
		75	*50*	*25*
714	Area, mi.^{2}	35	58	81
	Reservoir thickness, ft.	108	187	331
	Porosity	0.11	0.14	0.19
	Trap fill	0.25	0.39	0.61
409	Area, mi.^{2}	27	46	77
	Reservoir thickness, ft.	82	144	249
	Porosity	0.10	0.14	0.18
	Trap fill	0.21	0.34	0.53

has a multivariate lognormal distribution, ($\boldsymbol{\mu}$, $\boldsymbol{\Sigma}$), where $\boldsymbol{\Sigma}$ is positive definite. The mean and variance of X given $X = x$ can be estimated (see Appendix C).

The conditional probability distributions for the reservoir parameters were computed for each given pool size in the conceptual play. Examples of the values at the 75th, 50th, and 25th upper percentiles are given in Table 5.11.

A larger pool size has a larger variance for the area of closure, reservoir thickness, porosity, and trap fill than a smaller pool size. This phenomenon is the result of all the geological variables constrained by Equation 5.8. The conditional distributions of the same random variables for a given pool size partly overlap, reflecting the nature of the irregularities (e.g., small pool size with excellent porosity) and/or slight variation in random variables, such as porosity. This type of information can be used subsequently to calculate productivity. Estimated conditional pool area distributions can provide information for calculating the number of wells required for developing an undiscovered pool.

Constructing Probability Distributions

When estimating immature or conceptual plays, the probability distributions of geological random variables of a pool-size equation are needed to compute a pool-size distribution. Normally these probability distributions are constructed by geological judgment. In this section, guidelines for constructing probability distributions from geological information are outlined. For frontier plays, the assessment team should collect all relevant data and information from similar basins to address the following questions and concerns.

1. The first question that might arise in the case of frontier plays is: What is the probability that a play exists? This issue can be analyzed in terms of the presence or absence of factors such as source rocks, maturation, migration, and favorable reservoir facies. A marginal probability is applied to each factor to indicate the likelihood that the factor exists (geological factors are listed at the beginning of this chapter).
2. If a geological random variable in question has an extreme range of values, then its variance should be relatively large. On the other hand, if the values are uniform, then the variance should be small.
3. Remember that we do not have enough data to compute covariances between variables. However, positive or negative covariances are evident from geological data. Therefore, if the largest estimated pool size is not what we expect, the following questions should be addressed: Are the mean and variance of the pool-size distribution adequate? How much covariance exists?

Table 5.12. Format for Entry of Probability Distributions

Geological variable	*Unit of measurement*	*Probability in upper percentile*			
		1.0	*0.5*	*0.02/0.01*	*0.00*
Area of closure of pool	$mi.^2$ or km^2				
Net pay/no. of pay zones	m or ft./no.				
Reservoir or formation thickness	m or ft.				
Porosity	Decimal fraction				
Trap fill	Decimal fraction				
Favorable facies	Decimal fraction				
Water saturation	Decimal fraction				
Oil or gas saturation	Decimal fraction				
Shrinkage factor	Decimal fraction				
Formation volume factor	Decimal fraction				
Reservoir temperature	°Celsius or Fahrenheit				
Reservoir pressure	kPa or psi				
Recovery factor	Decimal fraction				

4. What is the value that just exceeds the maximum that the model can reasonably be expected to have? This value will be set at zero in the upper percentile of the probability distribution.
5. What is the largest possible value that the model can have? This will be set at the 99th or 98th upper percentile of the probability distribution.
6. What is the value that is exceeded by half the members of the population? This will be placed at the 50th percentile of the distribution.
7. What is the minimum value? This will be set at the 1st percentile.
8. In determining the geological factors that dictate the final accumulation of hydrocarbons, one might ask: What are the most unpredictable risk factors in this model?
9. The number of prospects can be obtained from anomalies showing closure on a structural contour map of time isochrons constructed from seismic data. However, some questions

Table 5.13. Format for Entry of Geological Factors and Their Marginal Probability

Geological factors	*Marginal probability*	*Play-level prospect*
Presence of closure		
Presence of reservoir facies		
Presence of porosity		
Adequate seal		
Adequate timing		
Adequate source		
Adequate maturation		
Adequate preservation		
Adequate recovery		
Adequate play conditions		
Adequate prospect conditions		

Table 5.14. Format for Entry of Number of Prospects and Pools

Geological variable	*Probability in upper percentile*		
	0.99	*0.5*	*0.0*
No. of prospects			
No. of pools			

remain unanswered: How many anomalies were not detected by the current orientation and density of seismic lines? What is the maximum number of prospects that could exist in this play? How many prospects would there be at a 50% chance? The answers to these questions provide us with information needed to construct probability distributions for the prospects. Other values at various upper percentiles can also be used.

For each probability distribution, values of the four upper percentiles (1.0, 0.5, 0.02 or 0.01, and 0.0) are the minimum requirement for constructing a distribution. The process commences by fitting a lognormal distribution to these four values and then generates all other upper percentiles. Assessors can either (1) enter the four upper percentiles and let the shape of a lognormal distribution generate other percentiles or (2) enter all percentiles and examine the difference between the input percentiles and the lognormal approximation.

Table 5.12 presents a sample format for tabulating a probability distribution. Samples for tabulating exploration risks (Table 5.13) and numbers of prospects and pools are also presented (Table 5.14).

6

Estimation Update and Feedback Procedures

> Far better an approximate answer to the right question, which is often vague, than an exact answer to the wrong question, which can always be made precise.
>
> —John W. Tukey

A basin or subsurface study, which is the first step in petroleum resource evaluation, requires the following types of data (see Table 6.1):

- *Reservoir data*—pool area, net pay, porosity, water saturation, oil or gas formation volume factor, in-place volume, recoverable oil volume or marketable gas volume, temperature, pressure, density, recovery factors, gas composition, discovery date, and other parameters (refer to Lee et al., 1999, Section 3.1.2).
- *Well data*—surface and bottom well locations; spud and completion dates; well elevation; history of status; formation drill and true depths; lithology; drill stem tests; core, gas, and fluid analyses; and mechanical logs.
- *Geochemical data*—types of source rocks, burial history, and maturation history.
- *Geophysical data*—prospect maps and seismic sections.

Well data are essential when we construct structural contour, isopach, lithofacies, porosity, and other types of maps. Geophysical data assist us when we compile number-of-prospect distributions and they provide information for risk analysis. The number of dry holes and

Table 6.1. Petroleum Resource Assessment Record Sheet

I. General Information

Country:	Geological province:
Basin:	Play:
Geologists:	Assessors:
Date of assessment: _____/_____/_____	
Date of completion of the sheet: _____/_____/_____	

II. Availability of Basic Information

A. Maps:	Source
Structural	
Isopach	
Facies	
Oil/gas pool locations	
Geophysical	
Cross-section	
Logs	
Other	
B. Comments on information availability:	
C. Level of knowledge concerning this play:	
No drilling and no seismic:	
No drilling but with seismic:	
Early stage—immaturely explored:	
Intermediate stage—fairly well explored:	
Late stage—maturely explored:	
Completely explored:	
D. Stratigraphy of [each] formation or pay:	
Formation name:	
Age:	Lithology:
Thickness:	Organic type:
Sedimentary environment:	
Heat flow:	Temperature gradient:
Oil window:	Gas window:
Pressure: Normal	Abnormal
Other comments:	
Indications of oil and/or gas:	

III. Statistics of the Play

A. Play area: mi.^{2} or km^2
Play area explored:
Play producing:
Play volume: mi.^{3} or km^3
Play volume explored:
Play volume producing:
B. Reservoir data:

	Oil			Gas		
	Minimum	Average	Maximum	Minimum	Average	Maximum
Pool area, ha						
Net pay, m						
Porosity						
Water saturation						
Depth, m						
Recovery factor						

C. Hydrocarbon volume (oil, 10^6m^3 or MMbbls; gas, 10^6m^3 or Bcf or Tcf):
In-place oil volume:
In-place gas volume:
Primary oil reserve:
Primary gas reserve:
Enhanced oil reserve:
Enhanced gas reserve:
Cumulative oil production:
Cumulative gas production:
D. Drilling history:
Number of wells penetrating the play:
Number of exploratory wells:
Number of exploratory wells interpreted as true test wells:
Number of development wells:
Mean recurrence time for a dry well:
Mean recurrence time for an oil well:
Mean recurrence time for an oil and gas well:
Exploration risk:
Ratio of producing area/play area:

the reasons why they fail provide information for estimating the marginal probability of each geological factor. Chronostratigraphic and organic maturation data are used to define a basin's burial and thermal history. All these data can be used to identify a play and its geographic boundaries. In addition, reservoir and well data retrieved within geographic play boundaries can provide the information needed to compile an exploration time series for evaluation of mature plays.

When there is not enough information to quantify every aspect needed, we can apply experience gained from other basins, or compile information from previous work. This type of compilation or comparative study can provide useful information in evaluating resources.

Procedure for Estimating Mature Plays

Step 1: Formulating a Play Definition and Its Geographic Boundary

A play has both geographic and stratigraphic limits; it is confined to a basin or part of a basin, to a structural unit or part of it, and also to one or more formations (Chapter 2). Figure 2.8 displays the areal extent of the Beaverhill Lake oil play example, as defined by the play definition and expressed on the map by the play boundary. By definition, all pools within a specific play form a natural geological population. The importance of a properly defined play is that it will correspond to a single statistical population and thus meet the statistical assumptions required for the proper operation or the evaluation processes. A mixed population resulting from an improperly defined play definition, for example, can adversely affect the quality of the final resource estimates.

Step 2: Compiling Play Data

Once a play is defined and the play boundary has been outlined on a map, all wells and pools within the formation(s) identified by the assessment team as being part of the play are retrieved from the PETRIMES information system (Lee et al., 1999). Each well or pool is then examined by the geologists to determine whether it is consistent with the play definition. If not, revisions of the play definition and boundary are made and retrieval is performed again.

Drill stem tests from exploratory wells within the polygon that conform to the play definition are examined to produce a complete exploration discovery time series. The exploration time series for the

Beaverhill Lake oil play is shown in Chapter 2 (Fig. 2.9). The horizontal axis indicates the discovery sequence for the wells drilled, and the gaps in the sequence represent the occurrence of dry holes. The upper vertical axis indicates the individual discovered in-place pool sizes, whereas the lower vertical axis indicates oil flow rates obtained from drill stem tests. These pool sizes and drill stem test recoveries are the basic input data required for resource assessment.

Oil or gas occurrences in a specific exploratory well can range in magnitude from a discovery of commercial size to the show of oil droplets or gas bubbles. Each occurrence can be considered, by definition, as a pool. In practice, however, an oil or gas accumulation is considered to be a pool only if it is of commercial value at the time of discovery. Imposing such a restricted definition on the underlying pool population has a severe impact on the validity of the resource estimate, because small pools in the population will be underrepresented and the amount of information needed to determine the total number of pools within a play will not be sufficient.

It is essential, therefore, to examine all possible potential pools that were not reported at the time of assessment. Although time-consuming and tedious, this extensive collecting of data is rewarding. It is much better to have an adequate data set for an assessment than to attempt to model the economic truncation problem from ill-defined statistical models. This is illustrated by the Beaverhill Lake example in Chapter 3.

Step 3: Validating Mixed Populations or Lognormal Assumptions

Having collected all the pool data for a play, two aspects must be validated: (1) the possible mixed populations and (2) the assumption of lognormality if LDSCV is used. A logarithmic probability plot such as that shown in Chapter 2 (Fig. 2.12) can be used to check whether these two attributes exist.

If the discoveries are thought of as a single population, then the empirical distribution function should exhibit an almost straight line on the plot. Also, if the discoveries obey a lognormal distribution, they should exhibit a straight line on the same plot. However, the statistical assumption required by the logarithmic probability plot is that the discoveries are a random sample from their population. This assumption, as we know, is not valid. The conclusion obtained from the plot is that there is no evidence to negate the hypothesis. Further Q–Q tests must be executed using the output derived by NDSCV.

Figure 2.12B is the logarithmic probability plot for the Keg River oil play in the Rainbow basin. Because the discoveries of the play show a fairly straight line on the plot, there is no evidence to negate either the single population or lognormal assumption hypothesis.

Step 4: Estimating Pool-Size Distribution

For mature plays, the pool-size distribution can be estimated using LDSCV and NDSCV. The log likelihood of both models suggests the total number of pools in the play. Therefore, the estimated β, μ, and σ^2 values can be obtained.

Step 5: Determining an Appropriate Probability Distribution

Having estimated the superpopulation pool-size distribution and the total number of pools, the NDSCV should be applied to the Q–Q plots to test the distribution assumptions. Figures 4.20, 4.22, 4.23, and 4.26 demonstrate that the lognormal assumption is adequate.

Step 6: Estimating Pool-Size-by-Rank

The pool-size-by-rank distribution for the Beaverhill Lake oil play was shown in Chapter 3 (Fig. 3.14A). Discovered pool sizes are represented by dots, whereas estimated pool sizes are indicated by vertical bars. Discovered pool sizes can be matched to specific estimated pool ranks during consultation with assessment team members.

Figure 3.14B shows pool-size-by-rank conditional on the match. By this we mean that undiscovered pool sizes have been constrained by the fact that their size ranges cannot be greater or less than any adjacent discovered pool. The expected value of each undiscovered pool is used in the economic analysis.

Step 7: Estimating Expected and Probable Play Potential

The remaining play potential can be estimated from the total number of pools and the pool-size distribution. Adding together the means of all undiscovered pool sizes yields the expected value of the remaining play potential distribution and is defined as the *expected potential.*

The expected value of the remaining potential is governed by individual pool sizes and the assigned pool ranks, both of which are determined by the geological play definition used and the quality of the data set for the discovered pools. If the discovered pool sizes are incorrectly estimated, appreciated, or depreciated, or if the rankings are altered, then the expected value of the remaining potential will be affected. Provided that the geology of a play is well understood and documented, the expected value should provide a reliable estimate of play potential.

It should be noted that given the possible truncation of the pool size data set, estimates of the resources in a play should not be considered as the ultimate resource for that play. The results of an assessment are for the pool size data set used, so the model only predicts the existence of undiscovered pools based on that data set. The probable play potential can also be derived (see Chapter 5).

Step 8: Computing Play Resource Distribution

One way to report the range of the play resource is to choose values at the 0.90 probability prediction interval (95th and 5th percentiles) of the play resource distribution.

Procedure for Estimating Conceptual Plays

Conceptual Plays from a Mature Basin

Unlike assessments for frontier basins, where the resources of conceptual plays may be determined from geological judgment, the number and size of conceptual plays that might exist in a mature basin can be estimated from NDSCV. Figure 6.1 shows the play resource discovery sequence for Devonian gas plays. This sequence was constructed after evaluating all mature Devonian plays and compiling their respective discovery dates.

Figure 6.2 shows the mature play-by-rank plot. Assuming that the Devonian mature plays belong to a single population, NDSCV can be used to estimate the size of each conceptual play and the total number of plays within the Devonian basin. As in the pool-size-by-rank plot, dots represent matched discovered plays and boxes represent plays yet to be discovered.

Conceptual Plays from a Frontier Basin

Step 1: Formulating Play Definitions

A conceptual play has both geographic and stratigraphic limits and is confined to a basin or part of a basin, or a structural unit or part of it, and is confined to one or more formations.

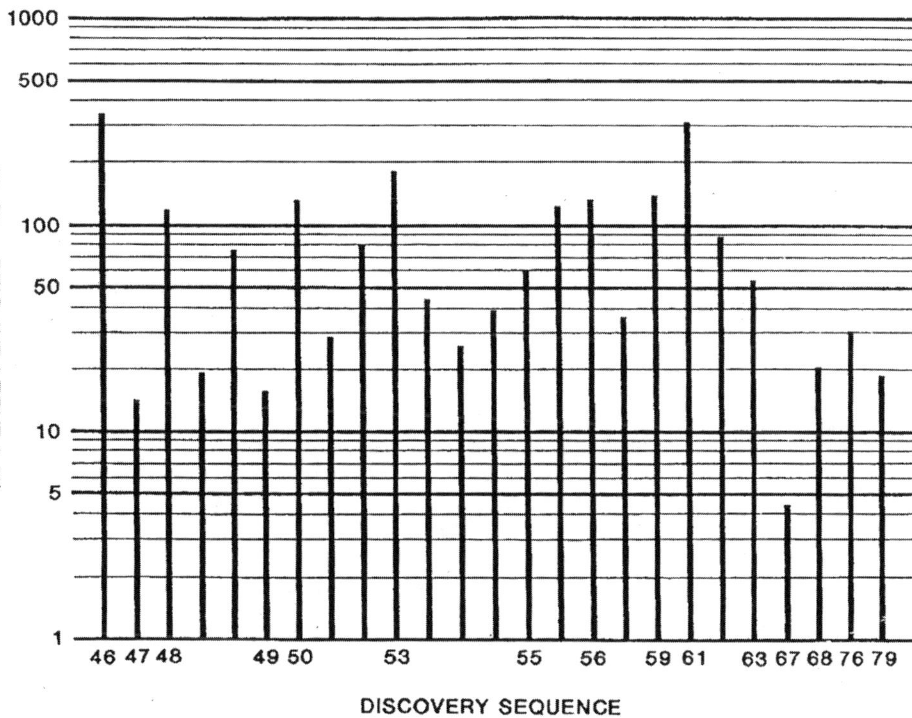

Figure 6.1. Discovery sequence plot for 25 Devonian gas plays of the Western Canada Sedimentary Basin by year of discovery (after Reinson et al., 1993).

Step 2: Estimating Pool-Size Distribution

Geological random variables, such as formation thickness, area of closure, porosity, trap fill, water saturation, and others, should be chosen and their probability distributions constructed from expert opinion and the results of comparative studies. Rules for constructing these types of distributions are given in Chapter 5. The distributions can be directly treated using the Monte Carlo method or they are approximated using the family of lognormal distributions for pool size calculation (Eq. 5.8). To verify lognormal approximations, both raw data and approximated distributions should be shown on the same plot.

Step 3: Estimating Number-of-Pools Distribution

The number-of-pools distribution can be computed by applying exploration risks and number-of-prospect distributions. For conceptual plays, both play- and prospect-level geological factors should be presented, whereas for mature plays only prospect-level geological factors are presented.

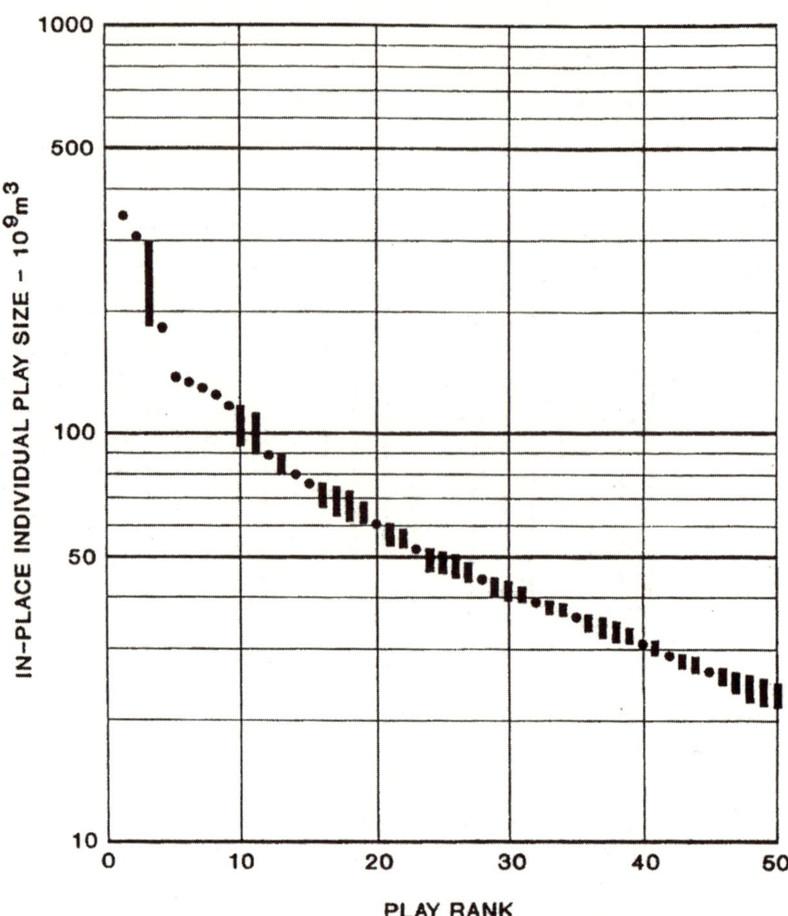

Figure 6.2. Play-size-by-rank plot for Devonian gas plays of the Western Canada Sedimentary Basin (after Reinson et al., 1993).

Step 4: Estimating Individual Pool-Size Distribution

The pool-size-by-rank can be computed when (1) N is a fixed value or (2) N is a random variable. Individual pool sizes can be further conditioned to the given pool ranks.

Step 5: Estimating Play Resource Distribution

The play resource distribution can be estimated using the estimated value of N and the pool-size distribution. A play potential distribution can be estimated by conditioning the play resource distribution to pool ranks.

Step 6: Estimating Other Reservoir Parameters

Reservoir parameters for each given pool size can be estimated. In addition, resource distributions for basin, geological province, or country can be obtained by summing all the potential from plays, basin, and geological provinces.

Update Procedure

A comprehensive and efficient annual update procedure is provided in PETRIMES, and changes are published as warranted. The update exercise is executed annually. Assessments of each play are updated in two steps, as follows:

1. New discoveries are examined to determine whether they lie inside or outside the play boundary.
2. The sizes of the new discoveries are then examined to determine whether they are consistent with those predicted.

If new discoveries do not agree with predictions, an update exercise is performed, revising play definitions for specific plays.

Feedback Procedure

Feedback is essential for any assessment. Figure 6.3 shows different levels of feedback through a petroleum assessment. It is extremely important for the validity of the assessment that it not be carried out in isolation. The geologists who defined the play and the assessors who make the prediction must all work together as a team. The strength of the PETRIMES approach to assessing hydrocarbon resources is derived from teamwork and constant feedback between the geological and statistical components of the assessment procedure. Separating these two components and "number crunching" the pool data can only lead to unreliable and potentially disastrous estimates. The feedback mechanisms are described in the following sections.

Can We Predict the Current Situation?

For mature plays, it is highly recommended that the discoveries be divided into two subsets: (1) to examine whether the second set can be

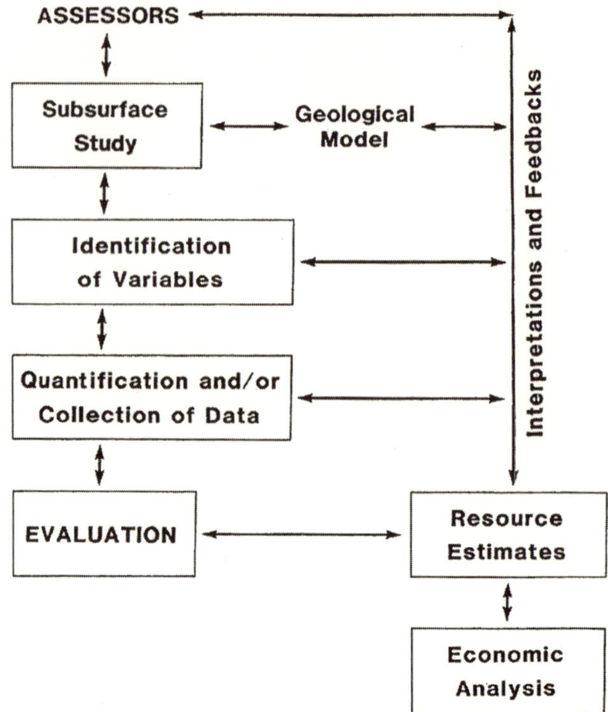

Figure 6.3. Diagram showing levels of feedback in the process of petroleum resource evaluation.

predicted from the first sample set and (2) to find an adequate prediction interval. The estimates must be validated by one or more of the following procedures:

- Comparison of the remaining largest pool size with geological models or exploration concepts
- Estimation of undiscovered pool sizes, matching discovered pool sizes and their ranks
- Retrospective study, as illustrated in Chapter 4

Has the Largest Pool Been Discovered?

For the Beaverhill Lake play example, in which the largest pool appears to have been discovered, geologists might ask: Is there a larger undiscovered pool? Or: What would the largest pool size be if the discovered largest pool is assumed to be the second largest in a play? Our method allows us to analyze these questions.

Take the Beaverhill Lake play as an example. Given that the largest discovered pool (211×10^6 m^3) is actually the second largest pool, then the predicted interval for the size of the largest possible pool ranges from 320×10^6 m^3 to 4129×10^6 m^3, which requires a pool area as large as the largest currently present pool. With this information, we can address the question: Have we overlooked the largest pool of this play? This type of feedback mechanism allows us to challenge underlying geological concepts or to validate our input data. It is one of the essential features of the evaluation system.

Pool Size Conditional on Play Resource

Individual pool size and number of pools can be estimated for a given play resource. This technique can be used as a feedback mechanism to resolve discrepancies between different estimates and to validate basic input factors, such as exploration risk, number of pools, and pool-size distribution.

Having computed the play resource distribution, one measure of the resource is the mean of the distribution. However, geologists might choose a value other than the mean of the distribution as a point estimate of the resource.

7

Other Assessment Methods—An Overview

"Would you tell me please which way I ought to go from here?"
"That depends a good deal on where you want to get to," said the Cat.

—Lewis Carroll

Resource evaluation procedures have evolved along distinct paths, involving a variety of statistical, geochemical, and geological approaches because of different types of data and various assumptions that have driven their development. Many methods have been developed so far, but only those methods that have been published and have significantly influenced subsequent development of evaluation procedures are discussed here. The purpose of this chapter is to present an overview of the principles of these methods and identify the direction of future research in this area. Methods discussed include the following:

- *Geological approach*—volumetric yield by analogy, basin classification
- *Geochemical approach*—petroleum systems, burial and thermal history
- *Statistical approach* (methods that were not discussed in previous chapters are discussed here)
- *Finite population methods*—Arps and Roberts', Bickel's, Kaufman's anchored, and Chen and Sinding–Larsen's geo-anchored

- *Superpopulation methods*—USGS log-geometric, Zipf's law, creaming, and Long's
- *The regression method*
- *The fractal method*

Specific data and assumptions can be applied to each of these methods. Some of the assumptions can be validated by the data whereas others cannot. These methods have their own merits and disadvantages.

Geological Approach

The geological approach has been used for the past several decades and is a qualitative method. This section discusses the volumetric yield method and the basin classification method.

Volumetric Yield by Analogous Basin Method

Volumetric yield using the analogous basin method was the earliest method of petroleum resource evaluation applied to frontier basins. It requires knowledge of the volume of a basin and its characteristics (e.g., tectonic, sedimentation, thermal generation, migration, and accumulation). Based on comparative studies, geologists are able to apply a hydrocarbon yield factor per unit volume (i.e., barrels of oil/cubic unit of sediment) from one known basin to an unknown basin with similar characteristics. Thus, for conceptual basins, this provides some information about the richness of an unknown basin. The advantages are the following:

1. It is suitable for the evaluation of conceptual basins.
2. It is easy to understand.
3. It combines geochemical data and/or experience from mature basins.

The disadvantages are:

1. The "mirror image" of a resource in one basin can be unreliable when applied to another basin.
2. The assessment obtained cannot be validated at the time of assessment.

3. The information provided is inadequate for economic study because it only generates an aggregate resource estimate. However, this method can still be applied to evaluate frontier basins when information is sparse.

Basin Classification Method

The relationship between basin characteristics and the abundance of resources has been studied by many researchers. The following two examples illustrate their relationship.

Klemme (1975, 1986) attempted to classify basins according to their tectonic history, morphology, and basin size. Suggesting that field size is influenced by the variation of a basin's morphology and size, he classified 65 producing basins of the world and computed the percentage of the largest five fields in terms of present-day reserves. The implication of Klemme's classifications is that basins have either concentrated or dispersed types of petroleum habitats. *Concentrated* means that most of the resources are distributed among the largest fields; *dispersed* means that most resources are distributed in many small fields. According to Klemme's classification, the percent of the present-day basin reserves contained in the five largest fields (Table 7.1) varies from one type of basin to another (Fig. 7.1).

Kingston et al. (1983a, b, 1985) collected data from about 600 identifiable sedimentary basins worldwide, classified them according to their tectonic history, and tallied the percent of productive basins (Fig. 7.2). This classification can provide information about the possible productivity of a basin in assessing conceptual plays. For example, if we know the tectonic history of an untested basin from which a play is being assessed, then the percentage of productive fields for that type of basin would permit us to determine the probability that it contains hydrocarbons.

Geochemical Approaches

The geochemical approach started in the early 1960s, flourished in the 1970s, and advanced in the 1980s. This section briefly discusses the principle of the petroleum system method and burial and thermal history.

Petroleum System or Geochemical Mass Balance Method

A petroleum system is defined as a stratigraphic unit that is a continuous body of rocks separated from surrounding rocks by regional

Table 7.1. Klemme's Basin Classification and Resources

Type area	Shape*	Profile†	Examples	Percent‡
I. Craton Interior Basins	C to E	S	Illinois	30
			Michigan	25
			Williston	24
			Denver	20
II. Continental Multicycle Basins	E to C	A	Piceance–Uinta	91
			Oriente	68
A. Craton margin			Wind River	63
			Big Horn	69
			Powder River	65
			Arkhoma	63
			Anadarko	50
			Sichuan	49
			Fort Worth	68
			Green River Overthrust	64
			Volga–Ural	38
			Alberta	26
			Permian	23
B. Craton-accreted basins	"Sag"	S	Erg Occidental	94
			Paris	76
			Erg Oriental	73
			Southern North Sea	73
			San Juan	72
			Timan–Pechora	67
			West Siberian	61
			Great Artesian	52
C. Crustal collision zone, convergent	E	A	Middle Caspian	42
Plate margin (closed)			East Venezuela–Trinidad	29
Plate margin (open)	E	A	Tampico	78
			Vera Cruz–Reforma–Campeche	70
			North Borneo	44
			Gulf Coast	16
III. Continental Rifted Basins	E	I	Cambay–Bombay	86
			Aquitaine	86
A. Craton and accreted zones			Reconcavo	84
			Suez	76
			Viking	65

(*Continues*)

Table 7.1. *(Continued)*

			Gippsland	64
			Central Graben	56
			Dnieper–Donetz	50
			Sirte	50
B. Rifted convergent margin	E	I	North Sumatra	95
			Maracaibo	92
			Middle Magdalena	92
			Vienna	88
			Santa Maria	87
			Cook Inlet	78
			Ventura–Santa Barbara	77
			Los Angeles	74
			Java Sea	74
			South Sumatra	72
			Central Sumatra	68
			Baku–Kura	54
			San Joaquin–Sacramento	49
C. Rifted passive margin	E	A	Cuanza	96
			Sergipe–Alagoas	80
			Campos	77
			N.W. Shelf	75
			Congo	73
			Gabon	65
IV. Delta Basins C to E		A	Mahakan	80
			Mackenzie	74
			Niger	11
			Mississippi	8

*Shape (areal): C, circular; E, elongate.
†Profile: A, asymmetrical; I, irregular; S, symmetrical.
‡Percent of present basin reserve BOE (barrels of oil equivalent) in five largest fields. Numbers from Klemme (1975, 1986).

barriers to lateral and vertical migration of liquids and gases. It must be a confined system in which the processes of petroleum generation, migration, and accumulation of oil and gas take place independently from surrounding sediments.

This method is based on the process of formation, migration, and accumulation of hydrocarbons. Spatial distribution of the total organic carbon percentage can be mapped if adequate samples are obtained and the organic carbon volume can be integrated for each source rock

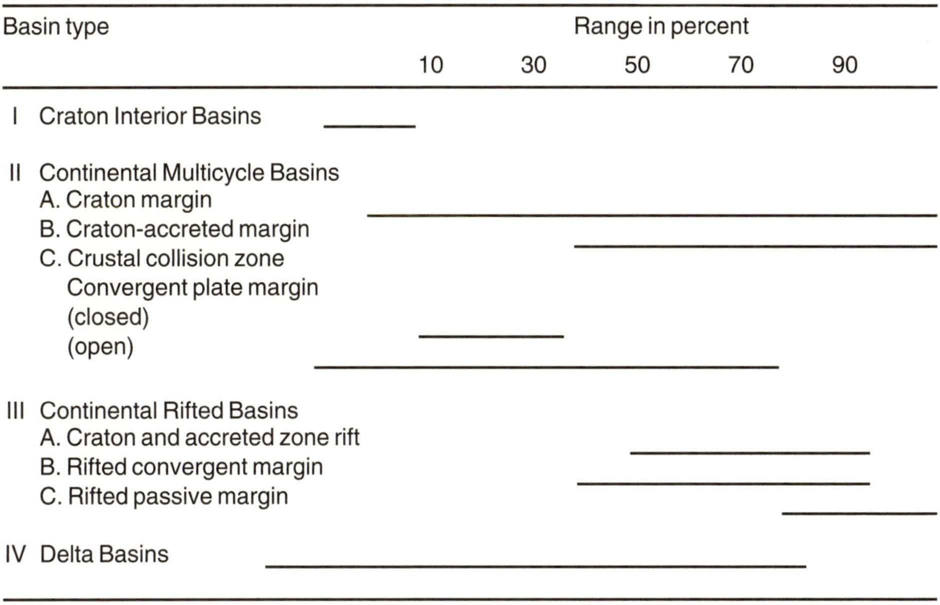

Basin type	Range in percent				
	10	30	50	70	90
I Craton Interior Basins					
II Continental Multicycle Basins					
A. Craton margin					
B. Craton-accreted margin					
C. Crustal collision zone					
Convergent plate margin					
(closed)					
(open)					
III Continental Rifted Basins					
A. Craton and accreted zone rift					
B. Rifted convergent margin					
C. Rifted passive margin					
IV Delta Basins					

Figure 7.1. Basins with various tectonic histories containing different petroleum resources.

bed. Degree of maturation can be delineated by geochemical data such as vitrinite reflectance, thermal alteration index, biomarkers, or burial history. The quantity of oil that can migrate from each source rock bed may be inferred from these studies by the following equation:

$$
\begin{aligned}
\text{oil generated} = \; & \text{Bulk rock volume of source rock} \\
& \times \text{Organic matter content by volume} \\
& \times \text{Genetic potential} \\
& \times \text{Fraction of oil in hydrocarbon yield} \\
& \times \text{Transformation ratio} \\
& \times \text{Volume increase on oil generation}
\end{aligned} \tag{7.1}
$$

The advantages of this equation are the following:

1. It is deterministic.
2. It provides a way of calculating an upper limit for the resources.
3. It is suitable for plays or basins.
4. It can be partially validated.

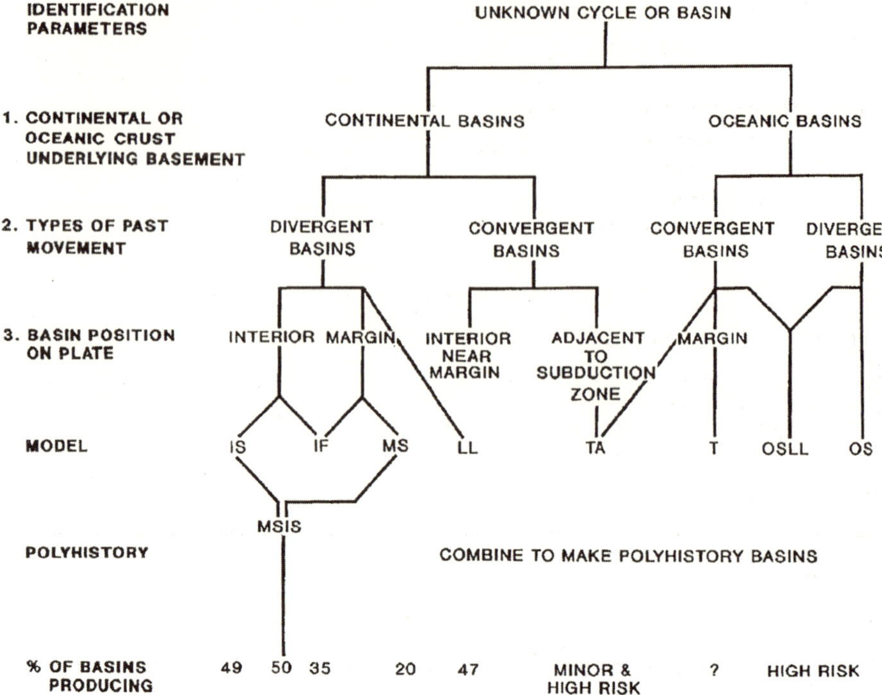

Figure 7.2. Basin classification according to tectonic history (after Kingston et al., 1983a, b, 1985) indicating the possibility of a basin containing hydrocarbons, based on its tectonic history. IF, interior fracture; IS, interior sag; LL, wrench; MS, margin sag; OS, oceanic sag; T, trench; TA, trench associated.

One of the unanswered questions is: How much oil and gas have been trapped in the basin? The severe drawback of this approach is that it is entirely deterministic (i.e., all geological processes are determined according to cause–effect relationships). The reliability of this approach depends on whether (1) the sample set adequately represents the basin or the play and (2) whether all relevant geological processes can be quantified in a deterministic way. Examples of this type of approach are given in Goff (1983) and Coustau et al. (1988).

Burial and Thermal History Modeling

The burial and thermal history of a hydrocarbon source bed or reservoir can be reconstructed if the following conditions are met:

1. A stratigraphic column displaying all formations and source beds and reservoirs has been identified.

2. The geological age of each source and reservoir bed is known.
3. The types of organic matter contained in each source bed are recognized.
4. The paleotectonic history of the stratigraphic column or paleo-heat flow of each source bed can be interpreted.
5. The thickness and lithology of each source bed can be estimated.

If the information relevant to the previous five conditions is available, a burial and thermal modeling process can proceed as follows:

1. Unpack the entire stratigraphic column into the thickness at the time of deposition.
2. Reconstruct the deposition and compaction of the stratigraphic column with consideration to erosional surfaces.
3. Record the thermal history of each source bed during the compaction process according to the principle of thermal dynamics given by the following equation:

$$\frac{dC_1}{dt} = K \times C_1 \tag{7.2}$$

where C_1 is the total amount of organic matter transformed into oil or gas, t is the time, and K is a coefficient defined as follows:

$$K = A \times e^{-E/RT}$$

The law of Arrhenius is a special case of the Boltzmann equation:

$$dC_1 = A \times e^{-E/RT} \times C_1 \times dt \tag{7.3}$$

where E is the molar activation energy (per mole) of the decomposition of certain types of organic matter into oil; A is a factor controlled by the decomposition condition, such as nature of the environment; C_1 is the amount of organic matter at time t; T is the absolute temperature; and R is the ideal gas constant.

The reconstruction process reveals (1) the geological time and depth of each source bed entering into and leaving the oil and gas windows, and (2) the current oil and gas windows.

Figure 7.3 shows an example from one of the basins situated in eastern China. The source bed is about 33.6 million years old, started to generate oil 26 million years ago, and entered the oil window 18.5 million

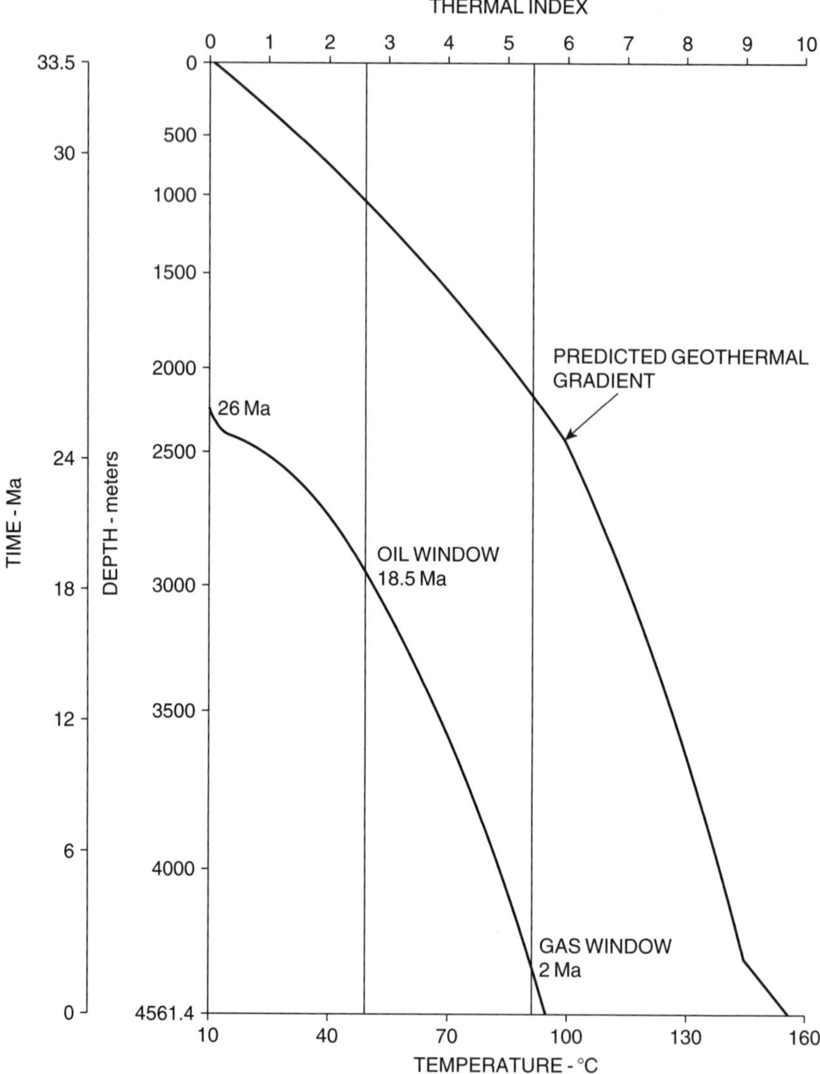

Figure 7.3. Burial history study for source bed ES2 from the Huang-Hua Basin of eastern China. The diagram indicates that the source bed started to generate oil 26 million years ago and entered into the oil and gas windows 18.5 million and 2 million years ago respectively.

years ago and the gas window 2 million years ago. The plot was based on the principle adopted by the DIAGEN program (du Rouchet, 1980; Lepoutré, 1986).

The advantage of this approach is that the method requires very little information yet can provide vital information about the maturity of source beds and the depths of the oil and gas windows. More comprehensive methods are described in Burrus et al. (1996).

Statistical Approaches

The merits of several discovery process methods for petroleum resource assessment were evaluated using discoveries from the Niagaran (Silurian) pinnacle reef play of northern Michigan as a benchmark data set for comparison (Gill, 1994; Lee and Gill, 1999). The tested methods included the USGS log-geometric method; the GSC PETRIMES methods, including LDSCV, NDSCV–empirical, nonparametric–lognormal, nonparametric–Pareto, and BDSCV methods; Arps and Roberts' method; Bickel, Nair, and Wang's nonparametric finite population; and Kaufman's anchored and Chen and Sinding–Larsen's geo-anchored methods (Table 7.2). The estimated number of fields varied by a factor of 3.7, but the estimated volume of resources varied by a factor of 1.6. The estimates are all fairly similar for the large field-size classes greater than 2 to 4 million barrels of oil equivalent (MMBOE). The main differences among the estimates are in the small fields less than 2 to 4 MMBOE.

This section reviews the advantages and disadvantages of the following statistical methods:

- The finite population approach
- The superpopulation approach
- The regression method
- The fractal method

Finite Population Methods

The Arps and Roberts Method

Arps and Roberts (1958) postulated that the probability of finding one more field with an area y in a basin for each additional wildcat to be drilled is proportional to (1) the magnitude of the area y of such fields and (2) the remaining number of undiscovered fields of that size. Therefore, the ultimate number of fields in any size class can be estimated from a negative exponential function as follows:

$$F_i(w) = F(\infty) \times \left(1 - \exp\left[\frac{-C_i \times W \times A_i}{B}\right]\right) \tag{7.4}$$

Table 7.2. Number of Fields and Total Resources of the Niagaran Pinnacle Reef Play, Michigan

Class interval, MMBOE	*No. of fields*								
	Discovered	*BDSCV*	*NDSCV empirical*	*Arps and Roberts method*	*LDSCV lognormal*	*Chen method*	*NDSCV Pareto*	*Bickel method*	*USGS method*
0.03–0.06	10	90	128	68	99	252	758	488	1333
0.06–0.12	20	97	123	83	106	260	545	528	850
0.12–0.25	42	146	150	150	154	264	396	590	541
0.25–0.50	71	168	148	188	181	265	288	545	345
0.50–1.0	101	170	142	193	166	202	209	441	202
1–2	138	146	152	180	138	179	152	358	142
2–4	85	85	86	92	85	91	86	142	85
4–8	59	59	59	59	59	59	59	72	59
8–16	9	9	9	9	9	9	9	9	9
16–32	3	3	3	3	3	3	3	3	3
Total no.	538	939–1008	1000	1042	1000	1584	2505	3176	3568
Total resource, MMBOE	1029	1145	1172	1252	1302	1341	1440	1855	1475

After Lee and Gill (1999).

where $F_i(w)$ is the number of fields found in the ith class by w exploratory wells, $F(\infty)$ is the total number of fields, W is the number of wells, C_i is the drilling efficiency for the ith size class, B is the basin area to be tested, and A_i is the average areal extent of fields in the ith size class.

The advantages of the method are twofold: (1) it is suitable for the evaluation of basins and (2) it provides quick estimates of the number and sizes of fields in a basin, and results can be used in economic research. The disadvantages are also twofold: (1) when estimating unknown population parameters, standard statistical methods do not apply for measuring uncertainty; and (2) the basin area to be tested, B, is difficult to estimate and is directly influenced by the number of fields to be estimated.

Take the Permian Basin of West Texas and southeastern New Mexico (Drew et al., 1980) as an example. The size class 10 has the following parameters: average areal extent of fields, 2.2 sq. mi.; cumulative exploratory wells through 1960, 14,243; and number of discoveries in size class 10 in the 0 to 5000-ft. interval through 1960, 59. The basin area is equal to 100,000 sq. mi., so the ultimate number of fields is 127. If the basin area were reduced by 50% (i.e., to 50,000 sq. mi.), the ultimate number of fields would be reduced to 83.

Bickel, Nair, and Wang's Method

Bickel's method (Bickel et al., 1992) is described as follows. Let $U = \{x_1, \ldots, x_N\}$ denote a finite population of N members and let Y_j be a characteristic associated with x_j, $j = 1, \ldots, N$. Let $S_n = (x_{i1}, \ldots, x_{in})$ be an ordered sample of size n that is selected successively without replacement and with probability proportional to some measure of size $\{w_1, \ldots, w_n\}$. More specifically,

$$P(\{x_{i1},\ldots,x_{in}\}) = \prod_{j=1}^{n} \frac{w_{i_1}}{\sum_{i=1}^{N} w_i - \sum_{k=1}^{j-1} w_{i_k}} \tag{7.5}$$

where $w_j = w(Y_j)$ is a positive function of the unknown population characteristic, and $w_{i_0} \equiv 0$. The likelihood function is as follows. Let

$$D(i) = \sum_{j=0}^{i-1} w(y_j), \quad i = 1,2,\ldots, n \tag{7.6}$$

with $w(y_0) \equiv 0$. The likelihood of $N = (N_1, \ldots, N_k)$ is obtained as

$$e^{L(N)} = \prod_{k=1}^{K} \frac{N_k!}{(N_k - n_k)!} \prod_{i=1}^{n} \frac{w(x_j)}{\sum_{r=-1}^{N} N_r w_r - D(i)} \tag{7.7}$$

where n is the number of discoveries of $x_1, \ldots, x_n$; N_k is the total number of pools in the kth class; n_k is the number of discovered pools in the kth class; $w(x_j)$ is a function of x such as x_j^{β}, where β is the exploration efficiency coefficient; and $D(i)$ is the discovery sequence as input.

This method simultaneously estimates the exploration efficiency, β, the total number of pools, N, and the number of undiscovered pools within each predefined size class. This method cannot be applied to cases when the sample size is too small.

Kaufman's Anchored Method

Kaufman (1986) established a variation on the Arps and Roberts method which stated that, given a well history H_w for which $X_j = n$, the probability that the $(w+1)$th well discovers a pool with area a is

$$\mathrm{P}(\tilde{Z}_{n+1} = a | a, H_w) = (N - n)p \tag{7.8}$$

where $p = ca/B$ given H_w, c is the exploration efficiency, a is the sum of all prospect areas to be tested, n is the number of discoveries at w wells drilled, and N is the total number of fields in the population.

It should be noted that parameter B is not the area of the basin. The value of B is the total area to be tested in the future. Therefore, it is equivalent to estimating the total number of prospects to be drilled and the sum of all prospect areas. Consequently, the expected number of discoveries made by the first w wells is

$$\bar{n}(w) \cong N(1 - e^{-caw/T})^{w} \tag{7.9}$$

If ca/T is small, then

$$\bar{n}(w) \cong N(1 - e^{-caw/T}) \tag{7.10}$$

which is the same as Arps and Roberts' equation (Eq. 7.4). It considers a finite population of N pools in a play, labeled $1, 2, \ldots, N$ and associated with a magnitude $x_j > 0$ to a field labeled j, $j = 1, \ldots, N$. Define

U = {1,2, . . . , N} and X = {$x_1, \ldots, x_N$}. A successive sampling model is a probability law that applies to the N! possible orderings in which elements of U can be observed. These probabilities depend on elements of x in the following fashion. Let ($i_1, \ldots, i_N$) be any ordering of all elements of U. The successive sampling is defined as follows:

$$\mathrm{P}(1,\ldots,N \mid x,\beta) = \prod_{j=1}^{N} \frac{x_j^{\beta}}{\left(x_j^{\beta} + \cdots + x_N^{\beta}\right)} \tag{7.11}$$

Given N and the discoveries S_n = {$x_1, \ldots, x_n$}, and let λ be a solution to

$$N = \sum_{j=1}^{n} \frac{1}{\left(1 - e^{-\lambda_1 x_j^{\beta}}\right)} \tag{7.12}$$

then

$$\hat{R}(N, \mathrm{S}_n) = \sum_{j=1}^{n} \frac{x_j}{\left(1 - e^{-\lambda_1 x_j^{\beta}}\right)} \tag{7.13}$$

is an approximately unbiased estimator of R. Given R, S_n and λ, a solution to

$$R = \sum_{j=1}^{n} \frac{x_j}{\left(1 - e^{-\lambda_2 x_j^{\beta}}\right)} \tag{7.14}$$

then

$$\hat{N}(R, \mathrm{S}_n) = \sum_{j=1}^{n} \frac{1}{\left(1 - e^{-\lambda_2 x_j^{\beta}}\right)} \tag{7.15}$$

is an approximately unbiased estimator of N. The exploration efficiency, β, can be estimated by other methods (e.g., LDSCV, NDSCV) and inserted into the equations as an exponent of the attribute, A.

This method is useful for testing geological concepts given N or R, particularly when geologists wish to know how many pools are required to make up a given resource inferred by judgment.

Chen and Sinding–Larsen's Geo-Anchored Method

Chen and Sinding–Larsen's geo-anchored method (Chen, 1993) has the same successive sampling property as Equation 7.5 and solves Equations 7.16 and 7.17 (Chen, 1993, Eqs. 3.25 and 3.26):

$$\hat{R} = \frac{\sum_{i=1}^{n} y_i}{\left[1 - \exp\left(-y_i^\beta \sum_{k=1}^{n} \frac{1}{\hat{T} - \sum_{l=0}^{k-1} y_l^\beta}\right)\right]}, \quad y_0 = 0 \tag{7.16}$$

$$\hat{N} = \sum_{i=1}^{n} \frac{1}{\left[1 - \exp\left(-y_i^\beta \sum_{k=1}^{n} \frac{1}{\hat{T} - \sum_{l=0}^{k-1} y_l^\beta}\right)\right]}, \quad y_0 = 0 \tag{7.17}$$

with $\hat{T}$ being a unique solution to Equation 7.18,

$$\hat{T} = \frac{\sum_{j=1}^{n} y_j^\beta}{\left[1 - \exp\left(-y_j^\beta \sum_{k=1}^{n} \frac{1}{\hat{T} - \sum_{l=0}^{k-1} y_l^\beta}\right)\right]}, \quad y_0 = 0 \tag{7.18}$$

where $\hat{T} = y_1^\beta + y_2^\beta + \cdots + y_N^\beta$, $\hat{N}$ is the estimated number of pools, $\hat{R}$ is the estimated resource, y_j is pool size, and n is the number of discoveries.

Superpopulation Methods

The PETRIMES method adopts the concept of the superpopulation approach and estimates the superpopulation distribution based on discovery process models, including the lognormal and nonparametric models. A number of other methods estimate the superpopulation parameters with varieties of estimation methods. We shall discuss them briefly.

USGS Log-Geometric Method

The USGS method entails a two-stage procedure, which combines the Arps and Roberts discovery process method (as described in this chapter) and the fitting of a log-geometric field-size distribution to the observed discoveries. The computational aspects are outlined in Drew (1990, pp. 147–171). The procedure, described by Drew and Schuenemeyer (1993), is as follows:

> The general form of the parent distribution was isolated by using a two-stage estimation procedure. In the first stage, the number of fields in each size class larger than the mode are estimated directly by using a discovery process model. In the second stage, the number of fields in size classes smaller than the mode are estimated by a technique based on inference. Then the two parts are joined together to construct the total field-size distribution. The technique used to estimate the number of fields in the size classes at and below the mode of the observed distribution depends on recognizing that this part of the underlying distribution is hidden behind the barrier of cost truncation. This barrier can be removed by using an inference gained from a study of the collective behavior of the truncation phenomenon across exploration plays and basins that have different cost/price regimes. Specifically, this inference is based upon the observation that the ratios of the estimated ultimate number of fields in successive size classes above the mode were, on average, constant. The underlying distribution of oil and gas fields estimated by this procedure is log-geometric in form. The essential component of this distributional form is that there are more fields occurring in each successively smaller field-size class. (p. 473)

The disadvantages include the following:

1. A constant ratio between the two adjacent size classes is difficult to validate. For example, the graphs shown by Schuenemeyer and Drew (1983, Fig. 4) display a random pattern (no trend), but can we conclude the ratios are constant because they exhibit random patterns?
2. The statistical assumption, log/geometric distribution, might not be valid.
3. Field sizes must be classified.
4. This method usually presents too large a number of small pools (Table 7.3).

Examples of this approach are presented in the paper by Schuenemeyer and Drew (1983).

Table 7.3. Comparison of the Estimations Derived by Zipf's Law, the Petroleum System Method, and the Discovery Process Methods

Methods	*Recoverable oil resource, Bbbls*
Zipf's law	Dispersed habitat with some undiscovered
Petroleum system	Oil generated = 88
PETRIMES	
Undiscovered	Middle Jurassic = 8.4–10.5
	Lower Jurassic = 2.0–2.3
Discovered	Upper Jurassic = 0.48
	Middle Jurassic = 9.88
	Lower Jurassic = 3.10
	Total = 13.46
Total resource	18.5–19.4

After Coustau et al. (1988).

Furthermore, Coustau (1981) adopted Zipf's law (Zipf, 1949) and stated that

$$S_m/S_n = (n/m)^k \tag{7.19}$$

where S_m is the pool size of rank m, S_n is the pool size of rank n, and k is a constant.

Taking $k = 1$ as an example, Equation 7.19 states that the largest pool size is twice as large as the rank 2 pool, and three times the size of the rank 3 pool, and so on. This implies that if the ratios between two adjacent ranked pools do not approximate the constant, then additional undiscovered pools might exist in size rank between the two. Coustau (1981) displayed pool-size-by-rank on a doubly logarithmic diagram. In this approach, the pools were arranged according to their descending order of size, and a rank was allocated to each of the pools. This suggested that if the lines declined with a gentle slope, then the play had a "dispersed habitat"; whereas if the lines declined with a steep slope, then the play had a "concentrated habitat." *Dispersed habitat* and *concentrated habitat* are terms defined by Klemme (1986). Comparisons between the methods of Zipf's law, geochemical mass balance, and the PETRIMES discovery process method were published by Coustau et al. (1988) and are listed in Table 7.3.

The Creaming Method

The creaming model (Meisner and Demirmen, 1981) makes use of a generally observed phenomenon that occurs in exploration provinces. This phenomenon, which may be referred to as *creaming,* is the diminishing effectiveness of exploration as it continues. The method assumes that the underlying pool-size distribution of a basin is lognormal and postulates that (1) the mean of log field size is a linear function of the corresponding exploratory well number and (2) the probability of success is a linear logistic function of the cumulative number of exploratory wells.

The creaming method is defined as follows: A discovered pool volume has a probability distribution with a density proportional to a power of its volume, and the density $v_i(x)$ of the ith wildcat well's discovery is proportional to

$$X^{Y_1+Y_2} f(x) \tag{7.20}$$

where Y_1 and Y_2 are the characteristics of the basin studied. Therefore, at the ith wildcat well, the discovered pool size has the following lognormal distribution:

$$\ln\left(x|\mu+\beta_1+\beta_2 i, \sigma^2\right) \tag{7.21}$$

where $\beta_1 = \mu + Y_1 \sigma^2$ and $\beta_2 = Y_2 \sigma^2$, and μ and σ^2 are the mean and variance of the lognormal superpopulation distribution respectively.

The creaming method is applicable to areas where discoveries are generally declining or constant. The estimates derived by the procedure are for short-term prediction only. Furthermore, the likelihood function of this method cannot be solved (Lorentziadis, 1991; Meisner and Demirmen, 1981), and the finite number of oil or gas fields is not captured by the creaming method.

Forman and Hinde (1985) extended this model by fitting a straight line to the plot of log field size versus discovery number and used extrapolations to indicate the likely size of future discoveries. However, this approach requires knowledge of the discovery order and can only predict average declining pool sizes.

Lorentziadis (1991) generalized the original creaming model by eliminating the lognormality assumption and obtaining a statistical solution for the model. Lorentziadis' model assumes that at the ith wildcat, the discovered size has a probability density $d_i(x)$ proportional to

$$d_i \propto X^{\beta_i/n} f(x) \tag{7.22}$$

where $f(x)$ is the superpopulation pool-size distribution of the population. Unlike the creaming method, the generalized model is solvable because the method does not require the parameter β_1, without which there is no loss of information. It is semiparametric in the sense that the superpopulation pool-size distribution consists of a parametric component and a nonparametric component.

Take, as an example, the Swan Hills–Kaybob South play of the Western Canada Sedimentary Basin (Reinson et al., 1993). Lorentziadis (1991) eliminated the first 41 failed wildcats and used the next 306 wildcats with 12 discoveries to predict the discoveries for the following 40 wildcats. The results of the lognormal and semiparametric approaches ranged from $117 \times 10^6\ m^3$ (median) to $4608 \times 10^6\ m^3$ (upper quartile) and $116 \times 10^6\ m^3$ (median) to $1741 \times 10^6\ m^3$ (upper quartile) respectively. The actual discovery is $473 \times 10^6\ m^3$ of in-place gas volume.

The Long Method

Long (1988) considered Kaufman's discovery process model as well as the effects of economic truncation and incomplete reporting of small pools. Assigning an economic truncation value to any given play is a difficult task, because whether a pool size is economic also depends on its size as well as its location.

Long claims that his method can account for size-biased data. Unfortunately, the Long method cannot estimate exploration efficiency. However, he suggests that the empirical relationship obtained by Forman and Hinde (1985) can be adopted here to estimate the value.

Before one can establish a fully satisfactory model that accounts for the truncation problem, one should incorporate all possible "pools" into the discovery sequence. The Long method was applied to the Bashaw data set and estimated that the Bashaw play contained 46 pools (Long, 1988, p. 119) instead of the 80 predicted by Lee and Wang (1985). By 1994, the Bashaw play discoveries numbered 75.

The Regression Method

If a basin or play has a long history of exploration and has a long time range and aggregated reserve data, then a regression method can be applied to the data and the total resources can be predicted by extrapolation. The method is defined as follows:

$$R_t = \frac{R_w}{\left(1 + e^{-\beta t}\right)} \tag{7.23}$$

where t is equal to time, R_w is the ultimate reserve, R_t is the reserve at time t, and β is the coefficient of exploration maturity.

The assumption is made that future additions of resources will increase according to this equation (Eq. 7.23). Examples of the application of the regression method to resource evaluation are shown in Figure 7.4, which demonstrates that these data sets can be approximated by the method. The advantages are that the method (1) fits data acquired over a long time range, as well as aggregated data; and (2) it is simple to apply. The disadvantages are (1) the statistical assumptions required by the regression might not be valid for future prediction from the current data, (2) the method is not suitable for predicting individual pool sizes, and (3) results from this kind of assessment are inadequate for economic study.

An example of this type of approach is found in Lee and Price (1991). A prediction of total petroleum resources by extrapolation of past exploration performance is a procedure commonly used in well-explored basins (Bettini, 1987). Finding rates over a long period of time are fitted by curves, and the area under the curve is integrated and interpreted as the ultimate reserve.

In addition, Dolton (1984) provides an example of this method by fitting the historical data from the Illinois Basin using both exponential and hyperbolic curves, which yielded different estimates. The exponential curve indicated that there were 38 MMbbls of recoverable oil, whereas the hyperbolic curve indicated that there were 115 MMbbls. This demonstrates that different mathematical functions used in the curve-fitting process can yield different estimates.

The Fractal Method

Lee and Lee (1994) demonstrate that distributions of some objects can be characterized by fractal properties. Fractal self-similarity is one of such properties incorporated in the power function, $y = x^{\beta}$. This function is versatile and has been used to describe other types of distributions worldwide, such as continental populations, areas of a continent, and river lengths.

Unlike Zipf's law, the fractal method does not require a constant ratio as stated in Equation 7.19, but does require knowledge of the size of the largest member in the population. Examples from the Leduc–Bashaw oil play and the Slave Point reef complexes–Cranberry gas play (Table 7.4) demonstrate that the fractal method has merit. From the predictions studied, we can assume that the few largest members

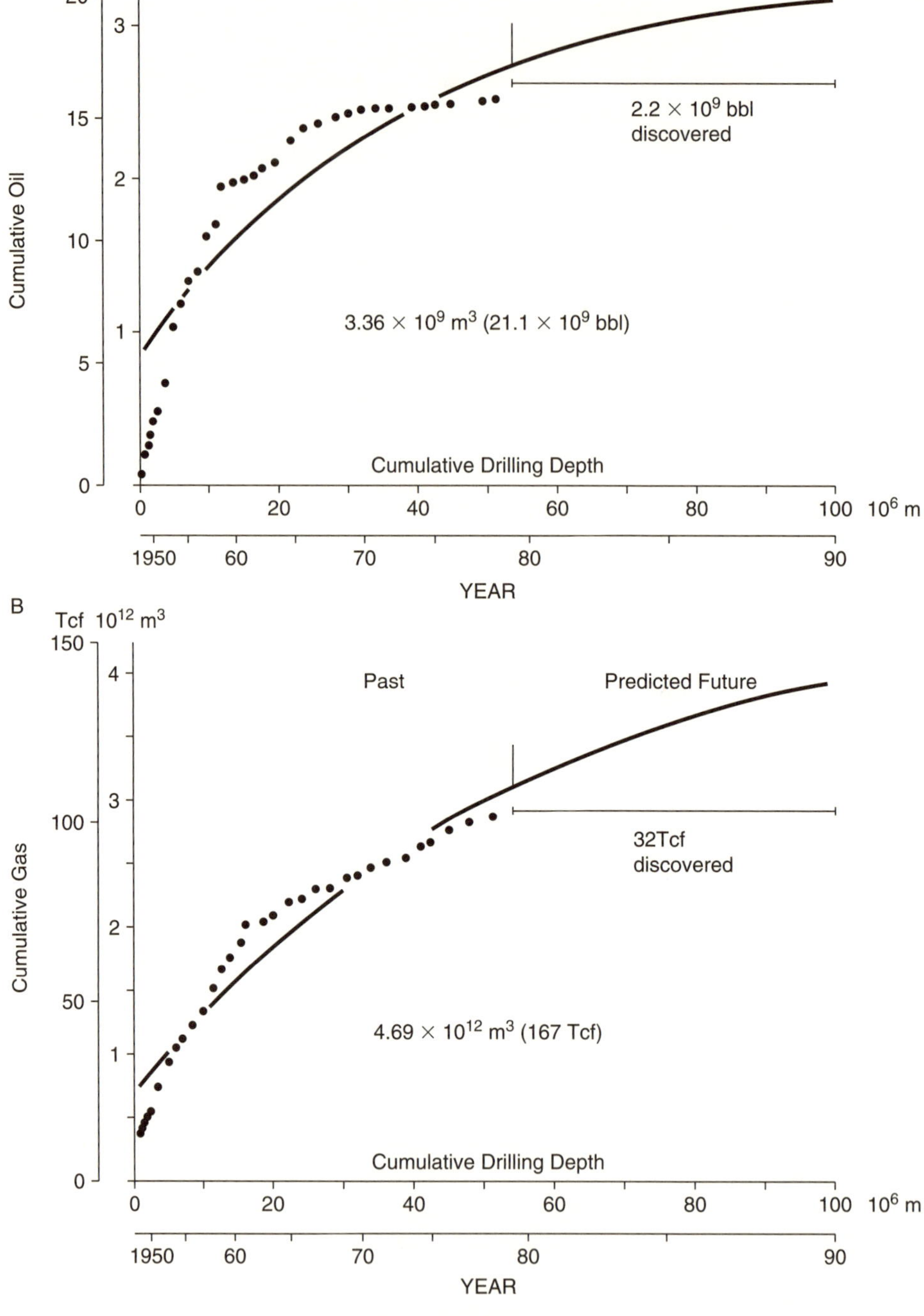

Figure 7.4. (A, B) Example regression models for petroleum resource evaluation (after Lee and Price, 1991). Dots represent reserves. The curve was derived by regression analysis. (A) Recoverable oil reserves. (B) Marketable gas reserves. By 1990, a total of 2.2 Bbbls of oil and 32 Tcf of gas had been discovered. Data from the Western Canada Sedimentary Basin.

Table 7.4. Comparisons between the Estimates Derived by the Superpopulation Approach and the Fractal Method

Pool rank	*Discovered, 10^6 m^3*	*Predicted by*	
		Superpopulation approach	*Fractal method*
A. Leduc–Bashaw oil play			
1	19.70	—	—
2	15.00	—	15.06
3	13.10	—	11.80
4	6.39	—	9.45
5	6.19	—	7.72
6	6.15	—	6.42
7	—	3.8–6.0	5.43
8	—	3.5–4.6	4.67
9	—	3.4–4.2	3.58
10	3.00	—	3.58
B. Slave Point complex reef–Cranberry gas play			
1	14,260	—	—
2	—	8113	8838
3	—	5444	5971
4	—	4051	4329
5	—	3218	3325
6	—	2667	2678
7	2276	—	2243
8	—	2035	1940
9	—	1833	1761
10	—	1661	1561

of the population follow a fractal distribution. Whether the remaining members of the population follow the same distribution has yet to be investigated.

Why does the fractal method yield estimates similar to those derived by the superpopulation method? The explanation, as we have demonstrated using Q–Q plots, is that pool-size distributions can also be described by a power normal distribution, which is identical to the power function used by Lee and Lee (1994).

8

Concluding Remarks

The procedure and steps of petroleum resource assessment involve a learning process that is characterized by an interactive loop between geological and statistical models and their feedback mechanisms. Geological models represent natural populations and are the basic units for petroleum resource evaluation. Statistical models include the superpopulation, finite population, and discovery process models that may be used for estimating the distributions for pool size and number of pools, and can be estimated from somewhat biased exploration data.

Methods for assessing petroleum resources have been developed using different geological perspectives. Each of them can be applied to a specific case. When we consider using a particular method, the following aspects should be examined:

- *Types of data required*—Some methods can only incorporate certain types of data; others can incorporate all data that are available.
- *Assumptions required*—We must study what specific assumptions should be made and what role they play in the process of estimation.

- *Types of estimates*—What types of estimates does the method provide (aggregate estimates vs. pool-size estimates)? Do the types of estimates fulfill our needs for economic analysis?
- *Feedback mechanisms*—What types of feedback mechanism does the method offer?

PETRIMES is based on a probabilistic framework that uses superpopulation and finite population concepts, discovery process models, and the optional use of lognormal distributions. The reasoning behind the application of discovery process models is that they offer the only known way to incorporate petroleum assessment fundamentals (i.e., realism) into the estimates. PETRIMES requires an exploration time series as basic input and can be applied to both mature and frontier petroleum resource evaluations.

Appendix A: Estimation of Superpopulation Parameters from a Successively Sampled Finite Population

Consider the superpopulation model in which unit values belonging to $\mathbf{Y}_N = (Y_1, \ldots, Y_N)$, a finite population of size N, are independent and identically distributed (i.i.d.) according to a cdf (cumulative distribution function), F. The prescribed sampling procedure is successive drawings without replacement, as follows: At each draw, the probability of selecting any particular unit is proportional to a weight function $\mathrm{w}(y)$ of its value if the unit remains in the population and is zero otherwise. Specifically, for a fixed sample of size $n \leq N$, the probability of observing an ordered sequence from the first to n is

$$\mathrm{P}\left[(i_1, \ldots, i_n) \mid \mathbf{y}_N\right] = \prod_{j=1}^{n} \frac{\mathrm{w}(y_{i_j})}{\sum_{i=1}^{N} \mathrm{w}(y_i) - \left[\mathrm{w}(y_{i_1}) + \cdots + \mathrm{w}(y_{i_{j-1}})\right]} \tag{A.1}$$

where $y_{i_0} = 0$ and $(i_1, \ldots, i_n)$ is an ordered sample of size n without replacement from $(1, \ldots, N)$. We assume that the weight function $\mathrm{w}(y)$ is positive and known, except for a finite set of unknown parameters. The population size N is also assumed known unless otherwise specified.

If the population of units is infinite, this sampling mechanism yields the selection-biased model studied by Cox (1969), Patil and Rao (1977, 1978), and more recently by Vardi (1982, 1985). In this case, the observations are i.i.d. with common cdf.

$$G(x) = \frac{\int_0^x w(y)\, dF(y)}{\int_0^\infty w(x)\, dF(x)}, \ x \geq 0 \tag{A.2}$$

When $w(y) = y$, we get the well-known length-biased model from an infinite population.

In petroleum resource evaluation of a hydrocarbon-bearing formation of a geological play, the finite population version of the selection-biased model plays an important role. This model provides a useful probabilistic framework for estimation of the pool-size distribution F of individual fields/pools while taking into account the "size-biased" phenomenon that often occurs in petroleum exploration (Arps and Roberts, 1958; Barouch and Kaufman, 1976, 1977; Kaufman et al., 1975). If the exploration history of a play has actually been dictated by the successive sampling mechanism shown in Equation A.1 with $w(y) = y$, the sample consisting of the first n discoveries is clearly not representative of the finite population. Indeed, the sample tends to be biased toward large sizes. Consequently, statistical methods based on random sampling will lead to erroneous inferences and generally provide overly optimistic predictions about sizes of undiscovered fields/pools in the play. On the other hand, if the size of pools had little or no impact on the order of discovery, a model based on sampling proportional to size without replacement yields pessimistic predictions.

Bloomfield et al. (1979), Smith and Ward (1981), and Lee and Wang (1985) consider the weight function in the form of $w(y) = y^{\beta}$ for a parameter β. This weight function includes the simple random sampling model ($\beta = 0$). The model with $\beta = 1$ was studied by Barouch and Kaufman (1976, 1977) when F is lognormal. In petroleum resource evaluation, the parameter β is known as the coefficient of discoverability and it is interpreted as a measure of the efficiency of the exploration process associated with the play. The larger the value of β, the more efficient the process.

In this appendix we shall assume that the superpopulation model distribution F is indexed by a vector of parameters, $\boldsymbol{\theta} = (\theta_1, \ldots, \theta_m)$, and each Y_i in $\mathbf{Y}_N$ has density $f(y|\boldsymbol{\theta})$. We consider the problem of maximum-likelihood estimation of $\boldsymbol{\theta}$ under the sampling model given in Equation A.1. The two-parameter lognormal distribution is of special interest, because petroleum geologists commonly use it for resource evaluation. Estimation of the coefficient of discoverability β and prediction of the population size N are also considered.

In the following section of Appendix A, we derive the likelihood function and propose a computational method for its evaluation. Maximum-likelihood estimations for $\boldsymbol{\theta}$ are next considered in the section "Maximum-Likelihood Estimation," together with examples of some specific forms of $f(y|\boldsymbol{\theta})$. The section "Inference for $\boldsymbol{\theta}$ and N" introduces a method of prediction for the population size N. In the final section, "Inference for the Weight Function," we consider the weight function $\mathrm{w}(y, \boldsymbol{\beta})$, where $\boldsymbol{\beta}$ is a vector of parameters, and examine the joint maximum-likelihood estimation of $\boldsymbol{\theta}$ and $\boldsymbol{\beta}$.

The Likelihood Function

Define $\mathbf{X}_N = (X_1, \ldots, X_N)$ as the vector of observations in order of occurrence so that X_j is the value observed for the jth draw. Upon relabeling the elements of $\mathbf{Y}_N = (X_1, \ldots, X_N)$ so that $X_j = Y_j$, $j = 1, 2, \ldots, n$, the probability of observing $\mathbf{x}_N = (x_1, \ldots, x_N)$, given $\mathbf{Y}_N = \mathbf{y}_N$, is

$$\mathrm{P}\left[(1, 2, \ldots, n) \middle| \mathbf{y}_N\right] = \prod_{j=1}^{n} \frac{\mathrm{w}(x_j)}{\mathrm{b}_j + \mathrm{w}(\mathbf{y}_{n+1}) + \cdots + \mathrm{w}(\mathbf{y}_N)} \tag{A.3}$$

where $\mathrm{b}_j = \mathrm{w}(x_j) + \cdots + \mathrm{w}(x_n)$. Multiplying Equation A.3 by the joint density of $\mathbf{Y}_N$ and integrating over the unobserved values $(Y_{n+1}, \ldots, Y_N)$ of $\mathbf{Y}_N$, the joint density of $X_1, \ldots, X_n$ is given as

$$\frac{N!}{(N-n)!} \prod_{j=1}^{n} f\left(x_j \middle| \boldsymbol{\theta}\right) \mathrm{E}\left[\prod_{j=1}^{n} \frac{\mathrm{w}(x_j)}{\mathrm{b}_j + \mathrm{w}(Y_{n+1}) + \cdots + \mathrm{w}(Y_N)} \middle| \boldsymbol{\theta}\right] \tag{A.4}$$

because $Y_1, \ldots, Y_N$ are i.i.d. and there are $N!/(N-n)!$ ordered samples of size n without replacement from a finite population of N units. Note that with $\mathbf{x}_N$ fixed and letting $N - n \to \infty$, the joint density shown in Equation A.4 approaches

$$\prod_{i=1}^{n} \frac{\mathrm{w}(x_i) f(x_i | \boldsymbol{\theta})}{\mathrm{E}[\mathrm{w}(Y_1) | \boldsymbol{\theta}]} \tag{A.5}$$

which is simply the likelihood of the infinite population selection-biased model of Equation A.2. The joint density given in Equation A.4 can be represented alternatively by

$$n!\binom{N}{n}\prod_{j=1}^{n} f(x_j|\boldsymbol{\theta})\prod_{j=1}^{n}\frac{\mathrm{w}(x_j)}{\mathrm{b}_j}\,\mathrm{E}\left[\prod_{j=1}^{n}\frac{\mathrm{b}_j}{\mathrm{b}_j+\mathrm{w}(Y_{n+1})+\cdots+\mathrm{w}(Y_N)}\Big|\,\boldsymbol{\theta}\right] \quad \text{(A.6)}$$

Now let $\varepsilon_1,\ldots,\varepsilon_N$ be independent and identically distributed exponential random variables with means equal to one and independent of $\mathbf{Y}_N$. Define λ_n as the sum of $\varepsilon_j/\mathrm{b}_j$, $j=1,2,\ldots,n$. Then the expectation term in Equation A.6 may be expressed as

$$\begin{aligned}&\mathrm{E}\left\{\mathrm{E}\left[\exp\left(-\boldsymbol{\Lambda}_n\{\mathrm{w}[Y_{n+1}]+\cdots+\mathrm{w}[Y_n]\}\right)\right]|\,\boldsymbol{\theta}\right\}\\ &=\mathrm{E}\left\{\mathrm{E}\left[\exp\left(-\boldsymbol{\Lambda}_n\,\mathrm{w}[Y_1]\right)\right]\,|\,\boldsymbol{\theta}\right\}^{N-n}\end{aligned} \quad \text{(A.7)}$$

Define $\rho_f(\lambda|\boldsymbol{\theta})$ as the Laplace transform of $\mathrm{w}(Y_1)$ with Y_1 distributed according to $f(y|\boldsymbol{\theta})$ and $d\mathrm{G}_n(\lambda)$ as the density of $\boldsymbol{\Lambda}_n$. Then combining Equations A.6 and A.7, the joint density of $X_1, X_2,\ldots, X_n$ is

$$n!\binom{N}{n}\prod_{j=1}^{n} f(x_j|\boldsymbol{\theta})\prod_{j=1}^{n}\frac{w(x_j)}{b_j}\int_0^{\infty}\left[\rho_f(\lambda|\boldsymbol{\theta})\right]^{N-n} d\mathrm{G}_n(\lambda) \quad \text{(A.8)}$$

Note that $\rho_f(\lambda|\boldsymbol{\theta})$ depends also upon the weight function $\mathrm{w}(y)$. According to Johnson and Kotz (1970, p. 222), $\boldsymbol{\Lambda}_n$ has a general gamma distribution with density given by

$$d\mathrm{G}_n(\lambda)=\sum_{l=1}^{n} C_l(\mathrm{b}_l e^{-\lambda b_l}),\quad \lambda>0 \quad \text{(A.9)}$$

where

$$C_l\prod_{i\neq l}\frac{\mathrm{b}_i}{\mathrm{b}_i-\mathrm{b}_l}=(-1)^{n-l}\prod_{i\neq l}\frac{\mathrm{b}_i}{|\mathrm{b}_i-\mathrm{b}_l|} \quad \text{(A.10)}$$

This density can be obtained by a partial fractions expansion of the Laplace transform of $\boldsymbol{\Lambda}_n$. It may be seen in Equations A.9 and A.10 that this density is a linear combination of exponential densities and is tied down at the origin, because $\sum C_l\mathrm{b}_l=0$. Also, it integrates to unity because $\sum C_l = 1$. This density is a data-dependent function through the partial sums $b_j=\sum_{i=j}^{n}\mathrm{w}(x_i)$ and is very sensitive to the order in which the observations are made.

From Equation A.8, the log likelihood of $\boldsymbol{\theta}$ given N, w(•), and data $\mathbf{x}_n$ is

$$\log L = \sum_{j=1}^{n} \log f(x_j|\boldsymbol{\theta}) + \log S(\boldsymbol{\theta}|\mathbf{x}_n) \quad \text{(A.11)}$$

where

$$S(\boldsymbol{\theta}|\mathbf{x}_n) = \int_0^\infty \rho_f(\lambda|\boldsymbol{\theta})^{N-n}\, dG_n(\lambda) \quad \text{(A.12)}$$

Note the following points:

1. If $w(y) \equiv 1$, the expectation in Equation A.4 is equal to $(N-n)!/N!$ and the joint density then reduces to the usual likelihood of n observations.
2. If the weight function is of the form $w(y) = y^{\beta}$ for an unknown parameter $\boldsymbol{\beta}$, the parameters $\boldsymbol{\theta}$ and $\boldsymbol{\beta}$ are not always identifiable in the infinite population selection-biased model shown in Equation A.10. For instance, if $f(y|\boldsymbol{\theta})$ is lognormal with parameters μ and σ^2, the sampling distribution is also lognormal with parameters $\mu + \boldsymbol{\beta}\sigma^2$ and σ^2. In the finite population case, it is shown by Equation A.8 that if $N = n$ (i.e., all units have been selected), the likelihood separates into two parts. The first part contains only information about $\boldsymbol{\theta}$, which is the usual likelihood of $\boldsymbol{\theta}$ given the data; the second part contains only information about $\boldsymbol{\beta}$, which is the probability of observing $X_1, \ldots, X_N$, in that order. Note also that the second part is just the marginal likelihood of $\boldsymbol{\beta}$ (Kalbfleisch and Prentice, 1973) under the Cox model for survival data (Cox, 1972) when there are no ties or censoring. When neither $N = n$ nor $\boldsymbol{\beta} = 0$, information about $\boldsymbol{\theta}$ and $\boldsymbol{\beta}$ is difficult to separate. There does not appear to be a partial likelihood decomposition (Cox, 1975) for $\boldsymbol{\beta}$ or $\boldsymbol{\theta}$. Principally, this is the result of the fact that unit values not included in the sample are unobservable. In the extreme case, with n fixed and $N - n \rightarrow \infty$, information about $\boldsymbol{\theta}$ and $\boldsymbol{\beta}$ is so mixed up that they cannot be separated.
3. The joint density of the observations derived by Barouch and Kaufman (1976) when $w(y) = y$ is the same as that shown in Equation A.8, except that we derived it in terms of the general gamma distribution. This form of Equation A.8 gives us the interpretation that the likelihood function for $\boldsymbol{\theta}$ and w consists of the usual individual likelihoods and an adjustment term that contains further but inseparable information about $\boldsymbol{\theta}$ and w.

In general, the integral $S(\boldsymbol{\theta}|\mathbf{x}_n)$ does not have a closed form for most of the commonly used superpopulation distributions, such as the lognormal distribution. Barouch and Kaufman (1977) computed a uniform asymptotic expansion for the density given in Equation A.8 when f_θ is lognormal, $\boldsymbol{\theta} = (\mu, \sigma)$, and $\mathrm{w}(y) = y$, then used it to approximate a likelihood function for μ, σ, and N given the data $\mathbf{x}_n$. Approximate conditional maximum-likelihood estimators for μ and σ^2, given N, were shown to be the unique maximizer of the uniform approximation to the likelihood. Although the uniform asymptotic approximation is valid for a wide range of possible parameter values of the lognormal density and for large $N-n$, its practical usage is somewhat limited. Estimates for the standard errors of the approximate conditional maximum-likelihood estimators are also not readily available.

Alternatively, the log-likelihood function of Equation A.11 may be numerically evaluated for each $\boldsymbol{\theta}$, given the observed data $\mathbf{x}_n$. At a casual glance at the integral in Equation A.12 with $d\mathrm{G}_n(\lambda)$ given by Equations A.9 and A.10, it appears that the most difficult part is the numerical evaluation of $\rho_f(\lambda|\boldsymbol{\theta})$ coupled with a suitable numerical quadrature routine. A closer examination (Barouch and Kaufman, 1977) reveals that the problem lies in the accurate evaluation of the general gamma density when λ is small. A direct calculation based on the density as defined by Equations A.9 and A.10 turns out to be numerically unfeasible unless the sample size n is small.

To see this, we first note that when λ is small

$$d\mathrm{G}_n(\lambda) = \prod_{j=1}^{n} \mathrm{b}_j \frac{\lambda^{n-1}}{(n-1)!} + 0(\lambda^n)$$

This follows from the fact that at $\lambda = 0$, the first $(n-2)$ derivatives of $d\mathrm{G}_n(\lambda)$ are zero, and the $(n-1)$st derivative is equal to $\Pi_{j=1}^{n} b_j$. Second, the coefficients C_l as defined by Equation A.10 can differ from the smallest to the largest by a very large factor, and they alternate in sign, so a large number of cancellations will occur near $\lambda = 0$. Because of rounding errors, the formulas in Equations A.9 and A.10 are practically useless for computing $d\mathrm{G}_n(\lambda)$ in the vicinity of the origin, $\lambda = 0$, where the most important contributions to the integral $S(\boldsymbol{\theta}|\mathbf{x}_n)$ occur.

When population values have no impact on the order in which the observations are made, that is, $\mathrm{w}(y) = 1$, we have

$$\mathrm{b}_j = n - j - 1, \quad j = 1, 2, \ldots, n$$

and

$$C_l = (-1)^{n-l} \prod_{i \neq l} \frac{n-i+1}{|l-i|}$$

so that

$$b_l\, C_l = (-1)^{n-l}\, n \binom{n-1}{l-1} \qquad \text{(A.13)}$$

and from Equation A.9, the general gamma density is given by

$$dG_n(\lambda) = ne^{-\lambda}(1-e^{-\lambda})^{n-1}, \qquad \lambda > 0 \qquad \text{(A.14)}$$

This density is seen to be the density of the largest order statistic of n i.i.d. unit exponential random variables. In this special case, the integral $S(\boldsymbol{\theta} \mid \mathbf{x}_n)$ reduces to

$$n \int_0^1 z^{N-n} (1-z)^{n-1}\, dz = \frac{1}{\binom{N}{n}}$$

which can be a very small number. On the other hand, a direct computation of the integral $S(\boldsymbol{\theta} \mid \mathbf{x}_n)$, with $dG_n(\lambda)$ as given by Equations A.9 and A.13, is numerically naive unless Equation A.13 can be represented with sufficient accuracy so that rounding errors will not accumulate in the cancellations of the sum in Equation A.9.

In the general case when $w(y) \neq 1$, these observations suggest that for the numerical approach, we must avoid calculating the partial fractions coefficients as defined by Equation A.10 in the evaluation of the general gamma density at each λ. One of the methods that achieve this end is the inverse Laplace transform. In our case, the Laplace transform of $dG_n(\lambda)$ is given as

$$l_n(s) = \prod_{j=1}^{n} \frac{b_j}{b_j + s} \qquad \text{(A.15)}$$

The inverse transform is given as

$$dG_n(\lambda) = (e^{a\lambda}/\pi) \int_0^\infty \{ \mathrm{Re}[l_n(s)] \cos w\lambda - \mathrm{Im}[l_n(s)] \sin w\lambda \}\, dw \qquad \text{(A.16)}$$

where $s = a + iw$, and a is any real number such that $l_n(s)$ is analytic for $\mathrm{Re}(s) > a$. Now, because $dG_n(\lambda)$ is of exponential order $-b_n$ (i.e., $|dG_n(\lambda)| \leq Me^{-b_n\lambda}$), Crump (1976) has shown that inverse transforms like Equation A.16 can be closely approximated over compact intervals using a Fourier series approximation. The approximation $d\tilde{G}_n(\lambda)$ on $(0, 2T)$ is given as

$$d\tilde{G}_n(\lambda)=(e^{a\lambda}/T)\left\{\frac{1}{2}l_n(a)+\sum_{k=1}^{\infty}[\mathrm{Re}(a+k\pi i/T)]\cos(k\pi\lambda/T)\right.$$
$$\left.-\mathrm{Im}[l_n(a+k\pi i/T)]\sin(k\pi\lambda/T)\right\} \tag{A.17}$$

where $dG_n(\lambda) = d\tilde{G}_n(\lambda)$ and the error E satisfies

$$\mathrm{E} \le \mathrm{M}e^{-\mathrm{b}_n\lambda}e^{-2T(a+b_n)}, \quad 0<\lambda<2T$$

It follows that by choosing a sufficiently larger than $-\mathrm{b}_n$, the error E can be made as small as desired. Of particular interest, Crump (1976) numerically demonstrated that for a sample size n as large as 200, the approximation formula given in Equation A.17 agrees with the special-case density in Equation A.14 to at least 10 significant figures.

Maximum-Likelihood Estimation

In this section we consider the inference for $\boldsymbol{\theta}$ when weight function $\mathrm{w}(y)$ and N are given. The maximum-likelihood estimator of $\boldsymbol{\theta}$, when it exists, can be obtained by the Newton–Raphson algorithm. Upon differentiating Equation A.11, the likelihood equations are

$$\sum_{j=1}^{n}\frac{\partial}{\partial\theta_r}\log f(x_j|\boldsymbol{\theta})+(N-n)\int_0^{\infty}\frac{\partial}{\partial\theta_r}\log\rho_f(\lambda|\boldsymbol{\theta})\,\xi(\lambda|\mathbf{x}_n,\boldsymbol{\theta})\,d\lambda=0 \tag{A.18}$$

where $r = 1, 2, \ldots, m$, and $\xi(\lambda|\mathbf{x}_n, \boldsymbol{\theta})$ is a data-dependent density function defined by

$$\xi(\lambda|\mathbf{x}_n, \boldsymbol{\theta}) = \rho_f(\lambda|\boldsymbol{0})^{N-n}\, dG_n(\lambda)/S(\boldsymbol{\theta}|\mathbf{x}_n), \quad \lambda \ge 0 \tag{A.19}$$

and

$$\frac{\partial}{\partial\theta_r}\log\rho_f(\lambda|\boldsymbol{\theta})=\frac{\mathrm{Cov}_{\boldsymbol{\theta}}\left\{\left[\frac{\partial}{\partial\theta_r}\log f(Y|\boldsymbol{\theta}),\ \exp[-\lambda\mathrm{w}(Y)]\right]\right\}}{\rho_f^{(\lambda|\boldsymbol{\theta})}} \tag{A.20}$$

If the maximum-likelihood estimate $\hat{\boldsymbol{\theta}}$ exists, it satisfies the likelihood equations in Equation A.18 and is the limit point of the iteration

$$\hat{\boldsymbol{\theta}}(v+1)=\hat{\theta}_n(v)+\mathbf{I}_0[\hat{\theta}_n^{(v)}]^{-1}\mathbf{U}[\hat{\theta}_n^{(v)}], \quad v=0,1$$

provided the initial estimator $\hat{\theta}_n^{(0)}$ is sufficiently close to $\hat{\theta}_n$, where $\mathbf{U}(\boldsymbol{\theta})$ is the $m \times 1$ vector of score functions defined by the left-hand side of Equation A.18 and $\mathbf{I}_0(\boldsymbol{\theta})$ is the second-derivative matrix of $-\log L(\boldsymbol{\theta})$. This has (r, s) entry

$$\mathbf{I}_{0,rs}(\boldsymbol{\theta}) = -\frac{\partial^2}{\partial\theta_r \, \partial\theta_s} \log L, \quad 1 \le r, \quad s \le m \tag{A.21}$$

To carry out the Newton–Raphson procedure, we need to calculate $(m+1)(m+2)/2$ double integrals for each iteration: one for the log likelihood, m for the score functions, and $m + \binom{m}{2}$ double integrals for the second-derivative matrix $\mathbf{I}_0(\boldsymbol{\theta})$. Under the successive sampling model of Equation A.1, the joint density of the remaining value $Y_{n+1}, \ldots, Y_N$ given data $\mathbf{x}_n$ is

$$f\left(y_{n+1}, \ldots, y_N \,|\, \mathbf{x}_n, \boldsymbol{\theta}\right) = \frac{\prod_{j=1}^{n} \mathrm{b}_j / \left[\mathrm{b}_j + \mathrm{w}(y_{n+1}) + \cdots + \mathrm{w}(y_N)\right] \prod_{k=n+1}^{N} f(y_k \,|\, \boldsymbol{\theta})}{S(\boldsymbol{\theta} \,|\, x_n)}$$

$$= \int_0^\infty \prod_{k=n+1}^{N} \frac{\exp\left[-\lambda \mathrm{w}(y_k)\right] f(y_k \,|\, \boldsymbol{\theta})}{\rho_f(\lambda \,|\, \boldsymbol{\theta})} \, \xi(\lambda \,|\, \mathbf{x}_n, \boldsymbol{\theta}) \, d\lambda$$

Define a density function, given λ and $\boldsymbol{\theta}$, as

$$\eta(a \,|\, \lambda, \boldsymbol{\theta}) = \frac{\exp[-\lambda \mathrm{w}(a)] f(a \,|\, \boldsymbol{\theta})}{\rho_f(\lambda \,|\, \boldsymbol{\theta})}, \quad a \ge 0 \tag{A.22}$$

Then

$$f\left(y_{n+1}, \ldots, y_N \,|\, \mathbf{x}_n, \boldsymbol{\theta}\right) = \int_0^\infty \prod_{k=n+1}^{N} \eta(y_k \,|\, \lambda, \boldsymbol{\theta}) \, \xi(\lambda \,|\, \mathbf{x}_n, \boldsymbol{\theta}) \, d\lambda \tag{A.23}$$

Note that this joint density is symmetrical in its arguments, and is a mixture of a product density. In particular, the conditional density of Y_{n+1}, given $\mathbf{x}_n$ and $\boldsymbol{\theta}$, at $Y_{n+1} = a$, is

$$f(a \,|\, \mathbf{x}_n, \boldsymbol{\theta}) = \int_0^\infty \eta(a \,|\, \lambda, \boldsymbol{\theta}) \, \xi(\lambda \,|\, \mathbf{x}_n, \boldsymbol{\theta}) \, d\lambda \tag{A.24}$$

Now, conditional on the data, let $\boldsymbol{\Lambda}$ follow the distribution shown in Equation A.19. Define A given $\boldsymbol{\Lambda} = \lambda$ as a random variable with density as given in Equation A.22. Then the conditional distribution of Y_{n+1}

given $\mathbf{x}_n$ is the marginal distribution of A in $(A, \mathbf{\Lambda})$, given the data. For a fixed λ, we have from Equations A.20 and A.22 that

$$\frac{\partial}{\partial\theta_r}\log \rho_f(\lambda|\boldsymbol{\theta}) = \mathrm{E}\left[\frac{\partial}{\partial\theta_r}\log f(A|\boldsymbol{\theta}) \mid \lambda, \boldsymbol{\theta}\right] \tag{A.25}$$

Therefore, the integral in the second term on the left-hand side of Equation A.18 can be written as

$$\begin{aligned}\mathrm{E}\left(\frac{\partial}{\partial\theta_r}\log \rho_r(\mathbf{\Lambda}|\boldsymbol{\theta})\mathbf{x}_n, \boldsymbol{\theta}\right) &= \mathrm{E}\left[\mathrm{E}\left(\frac{\partial}{\partial\theta_r}\log f(A|\boldsymbol{\theta})|\mathbf{\Lambda},\boldsymbol{\theta}|\,\mathbf{x}_n, \boldsymbol{\theta}\right)\right] \\ &= \mathrm{E}\left[\frac{\partial}{\partial\theta_r}\log f(A|\boldsymbol{\theta})|\,\mathbf{x}_n, \boldsymbol{\theta}\right]\end{aligned} \tag{A.26}$$

So, the likelihood equations in Equations A.18 and A.20 are simply given by

$$\mathrm{E}\left[\sum_{k=1}^{N}\frac{\partial}{\partial\theta_r}\log f(Y_k|\boldsymbol{\theta}) \mid \mathbf{x}_n, \boldsymbol{\theta}\right] = 0 \tag{A.27}$$

where $r = 1, 2, \ldots, m$, and maximum-likelihood estimates can be computed as solutions to Equation A.27. This may be interpreted to mean that if all the values in the finite population are known, then we can solve $\sum_{k=1}^{N}\frac{\partial}{\partial\theta_r}\log f(y_k|\boldsymbol{\theta}) = 0$ for the maximum-likelihood estimates. Because we do not know $\sum_{k=1}^{N}\frac{\partial}{\partial\theta_r}\log f(y_k|\boldsymbol{\theta})$, instead we shall solve its expectation given the data $\mathbf{x}_n$. This interpretation is precisely the idea behind the expectation–maximization (EM)algorithm that was introduced by Dempster et al. (1977) for computing maximum-likelihood estimates from incomplete data.

Barouch et al. (1983) illustrated the application of Equation A.27 when $f(y|\boldsymbol{\theta})$ is lognormal, and when sampling is proportional to size and without replacement. In the context of Dempster et al. (1977), the missing data are those values in the finite population that are not included in the sample. The complete-data log likelihood is

$$\log f(\mathbf{y}_N|\boldsymbol{\theta}) = \sum_{k=1}^{N}\log f(y_k|\boldsymbol{\theta})$$

Define for each pair $(\boldsymbol{\theta}, \boldsymbol{\theta}')$

$$\mathrm{Q}(\boldsymbol{\theta}|\boldsymbol{\theta}') = \mathrm{E}\left[\log f(\mathbf{Y}_N|\boldsymbol{\theta})|\,\mathbf{x}_n, \boldsymbol{\theta}'\right]$$

Then the EM iteration $\boldsymbol{\theta}^{(v)} \rightarrow \boldsymbol{\theta}^{(v+1)}$ is defined as follows:

E-step: Compute $Q[\boldsymbol{\theta} | \boldsymbol{\theta}^{(v)}]$.

M-step: Choose $\boldsymbol{\theta}^{(v+1)}$ to be a value of $\boldsymbol{\theta} \in E$, which maximizes $Q[\boldsymbol{\theta} | \boldsymbol{\theta}^{(v)}]$.

In the special case of the exponential families, the E-step and the M-step take special forms. In our problem, letting $\boldsymbol{\theta}^{(v)}$ be the current estimate of $\boldsymbol{\theta}$ after v iterations, the E-step is

$$Q\left[\boldsymbol{\theta} \middle| \boldsymbol{\theta}^{(v)}\right] = \sum_{j=1}^{n} \log f(x_j|\boldsymbol{\theta}) + (N-n)\, E\left[\log f(A|\boldsymbol{\theta}) \middle| \mathbf{x}_n, \boldsymbol{\theta}^{(v)}\right]$$

Then $\boldsymbol{\theta}^{(v+1)}$ of the M-step must satisfy $\frac{\partial}{\partial \theta_r} Q\left[\boldsymbol{\theta} \middle| \boldsymbol{\theta}^{(v)}\right] = 0$. That is,

$$\frac{n}{N} \bar{\mathbf{U}}_r \left[\boldsymbol{\theta}^{(v+1)} | \mathbf{x}_n\right] + \left(1 - \frac{n}{N}\right) E\left\{\frac{\partial}{\partial \theta_r} \log f\left[A | \boldsymbol{\theta}^{(v+1)} | \mathbf{x}_n, \boldsymbol{\theta}^{(v)}\right]\right\} = 0 \quad \text{(A.28)}$$

where $\bar{\mathbf{U}}_r(\boldsymbol{\theta}| \mathbf{x}_n) = \frac{1}{n} \sum_{j=1}^{n} \frac{\partial}{\partial \theta_r} \log f(x_j | \boldsymbol{\theta})$, the average of the incomplete-data score function for the rth component of $\boldsymbol{\theta}$. Now if $-\log f(\bullet|\boldsymbol{\theta})$ is convex, which is true for the exponential families, the M-step is equivalent to Equation A.28; hence, all limit points of any EM sequence $\{\boldsymbol{\theta}^{(v)}\}$ increase the likelihood equations in Equation A.18 or Equation A.27. Under fairly general conditions, Dempster et al. (1977) and Wu (1983) have shown that any EM sequence $\{\boldsymbol{\theta}^{(v)}\}$ increases the likelihood and will lead to a maximizer of the likelihood function. Also, if the likelihood function is unimodal and has only one stationary point, $\{\boldsymbol{\theta}^{(v)}\}$ converges to the unique maximizer $\hat{\theta}_n$, of the likelihood function.

To illustrate the EM algorithm, let us assume that the superpopulation distribution is lognormal with density given as

$$f(y|\boldsymbol{\theta}) = \frac{1}{y\sigma\sqrt{2\pi}} \exp\left[\frac{-(\log y - \mu)^2}{2\sigma^2}\right], \quad y > 0 \text{ and } \boldsymbol{\theta} = (\mu, \sigma)$$

The score functions are

$$\frac{\partial}{\partial \mu} \log f(y|\boldsymbol{\theta}) = (\log y - \mu)/\sigma^2 \quad \text{(A.29)}$$

$$\frac{\partial}{\partial \mu} \log f(y|\boldsymbol{\theta}) = \left[(\log y - \mu)^2 - \sigma^2 \right] / \sigma^2 \tag{A.30}$$

Define $\hat{\mu}_0 = \sum_{i=1}^{n} \log x_j / n$ and $\hat{\sigma}^2(\mu) = \sum_{j=1}^{n} (\log x_j - \mu)^2 / n$. Let $\{\boldsymbol{\theta}^{(v)}\}$ denote the current estimate of $\boldsymbol{\theta}$. Then the EM iteration $\boldsymbol{\theta}^{(v)} \rightarrow \boldsymbol{\theta}^{(v+1)}$ from Equation A.28 is given as

$$\mu^{(v+1)} = \frac{n}{N} \hat{\mu}_0 + \left(1 - \frac{n}{N}\right) \mathrm{E}\left[(\log A - \mu^{(v+1)})^2 \middle| \mathbf{x}_n, \boldsymbol{\theta}^{(v)} \right] \tag{A.31}$$

$$\sigma^{2(v+1)} = \frac{n}{N} \hat{\sigma}^2 (\mu^{(v+1)}) + \left(1 - \frac{n}{N}\right) \mathrm{E}\left[(\log A - \mu^{(v+1)})^2 \middle| \mathbf{x}_n, \boldsymbol{\theta}^{(v)} \right] \tag{A.32}$$

Given $\{\boldsymbol{\theta}^{(v)}\}$, we compute $\mu^{(v+1)}$ and substitute it into Equation A.32 to get the next estimate of σ^2, and repeat this procedure until either the log-likelihood function of Equation A.11 stops improving or the absolute difference between $\{\boldsymbol{\theta}^{(v)}\}$ and $\boldsymbol{\theta}^{(v+1)}$ is sufficiently small. To carry out this program, we need to calculate the conditional expectations.

Let $\boldsymbol{\theta}' = (\mu', \sigma')$. Then the conditional expectations of $\log A$ and $(\log A - \mu)^2$ given the data $\mathbf{x}_n$ and $\boldsymbol{\theta}'$ are

$$\mathrm{E}\left[\log A | \mathbf{x}_n, \boldsymbol{\theta}'\right] = \int_0^\infty \mathrm{E}\left[\log A | \lambda, \boldsymbol{\theta}'\right] \xi(\lambda | \mathbf{x}_n, \boldsymbol{\theta}')\, d\lambda \tag{A.33}$$

$$\mathrm{E}\left[(\log A - \mu)^2 | \mathbf{x}_n, \boldsymbol{\theta}'\right] = \int_0^\infty \mathrm{E}\left[(\log A - \mu)^2 | \lambda, \boldsymbol{\theta}'\right] \xi(\lambda | \mathbf{x}_n, \boldsymbol{\theta}')\, d\lambda \tag{A.34}$$

where $\xi(\bullet|\mathbf{x}_n, \boldsymbol{\theta}')$ is given by Equation A.19 and the conditional expectations inside the integrals are taken with respect to Equation A.22. Further manipulation, by using Equations A.25, A.29, and A.30, yields

$$\mathrm{E}\left[\log A | \lambda, \boldsymbol{\theta}'\right] = \mu' + \sigma^{2'} \frac{\partial}{\partial \mu'} \log \rho_f(\lambda | \boldsymbol{\theta}') \tag{A.35}$$

$$\begin{aligned} \mathrm{E}\left[(\log A - \mu)^2 | \lambda, \boldsymbol{\theta}'\right] &= (\mu' - \mu)^2 \\ &+ \sigma^{2'} \left[1 + 2(\mu' - \mu) \frac{\partial}{\partial \mu'} \log \rho_f(\lambda | \boldsymbol{\theta}') + \sigma' \frac{\partial}{\partial \sigma'} \log \rho_f(\lambda | \boldsymbol{\theta}') \right] \end{aligned} \tag{A.36}$$

Define $\phi(z)$ as the standard normal density. The Laplace transform and its partial derivatives with respect to μ and σ, in that order, are given as

$$\rho_f(\lambda|\boldsymbol{\theta}) = \int_{-\infty}^{\infty} \exp\left\{-\lambda w\left[\exp(\mu+\sigma z)\right]\right\}\phi(z)\,dz \qquad \text{(A.37)}$$

$$\frac{\partial}{\partial\mu}\rho_f(\lambda|\boldsymbol{\theta}) = \int_{-\infty}^{\infty} \varphi_\lambda(z|\boldsymbol{\theta})\,\varphi(z)dz,\ \frac{\partial}{\partial\sigma}\rho_f(\lambda|\boldsymbol{\theta}) = \int_{-\infty}^{\infty} \varphi_\lambda(z|\boldsymbol{\theta})\,z\phi(z)dz \qquad \text{(A.38)}$$

where $\phi_\lambda(z|\boldsymbol{\theta}) = -\lambda w\left[\exp(\mu+\sigma z)\right]\exp\left\{\mu+\sigma z - \lambda w\left[\exp(\mu+\sigma z)\right]\right\}$

and $w(y) = \dfrac{d}{dy}w(y)$.

For another illustration, let us consider the two-parameter gamma distribution with probability density function

$$f(y|\boldsymbol{\theta}) = \lambda^\alpha/\Gamma(\alpha)y^{\alpha-1}e^{-\lambda y}, \quad y>0 \quad \text{and} \quad \boldsymbol{\theta} = (\lambda, \alpha)$$

Given $\boldsymbol{\theta}^{(v)}$ as the current estimate of $\boldsymbol{\theta}$, it is easy to check that the M-step satisfies the equations

$$\alpha/\lambda = C_1\left[\mathbf{x}_n, \boldsymbol{\theta}^{(v)}\right]$$

$$\phi(\alpha) - \log\lambda = \log C_2\left[\mathbf{x}_n, \boldsymbol{\theta}^{(v)}\right]$$

where

$$C_1\left[\mathbf{x}_n, \boldsymbol{\theta}^{(v)}\right] = \frac{n}{N}\bar{x} + \left(1-\frac{n}{N}\right)\mathrm{E}\left[A\middle|\mathbf{x}_n, \boldsymbol{\theta}^{(v)}\right] \qquad \text{(A.39)}$$

$$C_2\left[\mathbf{x}_n, \boldsymbol{\theta}^{(v)}\right] = \exp\left\{\frac{n}{N}\hat{\mu}_0 + \left(1-\frac{n}{N}\right)\mathrm{E}\left[\log A\middle|\mathbf{x}_n, \boldsymbol{\theta}^{(v)}\right]\right\} \qquad \text{(A.40)}$$

and $\bar{x} = \sum_{j=1}^{n} x_j/n$, $\hat{\mu}_0 = \sum_{j=1}^{n}\log x_j/n$, and $\phi(x)$ is the digamma function. By Jensen's inequality, note that $C_2(\mathbf{x}_n, \boldsymbol{\theta}) < C_1(\mathbf{x}_n, \boldsymbol{\theta})$ for every $\boldsymbol{\theta}$ and $\mathbf{x}_n$. Therefore, the EM iteration $\boldsymbol{\theta}^{(v)} \to \boldsymbol{\theta}^{(v+1)}$ is given by the following two steps:

Step 1. Determine $\alpha^{(v+1)}$ as the solution of the equation

$$\log\alpha - \phi(\alpha) = C_0\left[\mathbf{x}_n, \boldsymbol{\theta}^{(v)}\right] \equiv \log\left\{C_1\left[\mathbf{x}_n, \boldsymbol{\theta}^{(v)}\right]/C_2\left[\mathbf{x}_n, \boldsymbol{\theta}^{(v)}\right]\right\} \qquad \text{(A.41)}$$

Step 2. Compute

$$\lambda^{(v+1)} = \alpha^{(v+1)}/C_1\left[\mathbf{x}_n, \boldsymbol{\theta}^{(v)}\right] \qquad \text{(A.42)}$$

Note that if $N = n$, one iteration of steps 1 and 2 solves the usual likelihood equations from the complete data. There are various methods for solving the root of Equation A.41. The trigamma function $\rho'(\alpha)$ is required if Newton's method is used. A very close approximation to $\alpha^{(v+1)}$ is given by the empirically determined formulas (see Johnson and Kotz, 1970, p. 189)

$$\alpha^{(v+1)} \approx C_0^{-1}(0.5000876 + 0.1648852\, C_0 - 0.0544274\, C_0^2) \quad 0 < C_0 < 0.5772$$

$$\alpha^{(v+1)} \approx C_0^{-1}(17.79728 + 11.968477\, C_0 - C_0^2)^{-1}$$
$$\times (8.898919 + 9.059950\, C_0 + 0.9775373\, C_0^2) \qquad 0 < C_0 < 0.5772$$

where $C_0 = C_0[\mathbf{x}_n, \boldsymbol{\theta}^{(v)}]$. The conditional expectations of A and $\log A$ given the data $\mathbf{x}_n$ and $\boldsymbol{\theta}^{(v)}$ can be obtained in a similar manner as in the lognormal case. When the weight function is given by $\mathrm{w}(y) = y$ (i.e., sampling proportional to magnitude),

$$\mathrm{E}\left[A \middle| \mathbf{x}_n, \boldsymbol{\theta}'\right] = \frac{\alpha'}{\lambda'}\left\{1 - \int_0^\infty \left[(t/\lambda')/(1+t/\lambda')^{\alpha'+1}\right]\, \xi(t|\mathbf{x}_n, \boldsymbol{\theta}')\, dt\right\} \tag{A.43}$$

$$\mathrm{E}\left[\log A \middle| \mathbf{x}_n, \boldsymbol{\theta}'\right] = \rho(\alpha')$$
$$= \log \lambda' - \int_0^\infty \left[\log(1+t/\lambda')/(1+t/\lambda')^{\alpha'}\right] \xi(t|\mathbf{x}_n, \boldsymbol{\theta}')\, dt \tag{A.44}$$

$$\xi(t|\mathbf{x}_n, \boldsymbol{\theta}') = \frac{(1+t/\lambda')^{-\alpha'(N-n)}\, d\mathrm{G}_n(t)}{\int_0^\infty (1+t/\lambda')^{-\alpha'(N-n)}\, d\mathrm{G}_n(t)} \tag{A.45}$$

In the previous two examples, we see that the EM algorithms are based on the complete-data sufficient statistics. This is not surprising, because in terms of natural parameters, both the lognormal and gamma distributions have the regular exponential-family form

$$f(y|\boldsymbol{\theta}) = \mathrm{b}(y)\exp\{t(y)^T \boldsymbol{\theta}\}/a(\boldsymbol{\theta}) \tag{A.46}$$

where $\boldsymbol{\theta}$ lies in an m-dimensional convex set Ξ such that Equation A.46 is a density for all $\boldsymbol{\theta} \in E$ and $t(y)$ is an $m \times 1$ vector of complete-data sufficient statistics. In this situation, the EM iteration $\boldsymbol{\theta}^{(v)} \rightarrow \boldsymbol{\theta}^{(v+1)}$ for our problem takes on the following form:

E-step: Compute $t^{(v)} = \frac{n}{N}\bar{t}(\mathbf{x}_n) + \left(1 - \frac{n}{N}\right)\mathrm{E}\left[t(A)\middle| \mathbf{x}_n, \boldsymbol{\theta}^{(v)}\right]$.

M-step: Solve $\boldsymbol{\theta}^{(\nu+1)}$ as the solution of the equation $\mathrm{E}[t(A)|\boldsymbol{\theta}] = \mathrm{t}^{(\nu)}$, where

$$\bar{t}(\mathbf{x}_n) = 1\Big/ n\sum_{j=1}^{n} t(x_j),\ \mathrm{E}\big[t(A)|\boldsymbol{\theta}\big] = \left[\frac{\partial}{\partial\theta_1}\log a(\boldsymbol{\theta}), \ldots, \frac{\partial}{\partial\theta_r}\log a(\boldsymbol{\theta})\right]^T,$$

and

$$\mathrm{E}\left[t(A)\big|\mathbf{x}_n, \boldsymbol{\theta}^{(\nu)}\right] = \int_0^\infty \mathrm{E}\left[t(A)\big|\lambda, \boldsymbol{\theta}^{(\nu)}\right]\xi\left[\lambda\big|\mathbf{x}_n, \boldsymbol{\theta}^{(\nu)}\right]d\lambda$$

This form of the EM algorithm is equivalent to Equation A.28 when the superpopulation model is a regular exponential family. According to Dempster et al. (1977), $\{\boldsymbol{\theta}^{(\nu)}\}$ will converge to some $\boldsymbol{\theta}^*$ in the closure of Ξ. The limiting $\boldsymbol{\theta}^*$ will occur at a local, if not global, maximum of the log likelihood given in Equation A.11, unless the observed Fisher information matrix is negative definite at $\boldsymbol{\theta}^*$. From the solution of Equation A.27, the (r, s)th entry of the second-derivative matrix of $\log L$ is

$$\begin{aligned}\frac{\partial^2}{\partial\theta_s\,\partial\theta_r}\log L = {}& \mathrm{E}\left[\sum_{k=1}^{N}\frac{\partial^2}{\partial\theta_s\,\partial\theta_r}\log f\left(Y_k|\boldsymbol{\theta}\right)\right]\\ &+(N-n)\mathrm{E}\left[\frac{\partial}{\partial\theta_r}\log f\left(A|\boldsymbol{\theta}\right)\frac{\partial}{\partial\theta_s}\log f\left(A|\,\mathbf{x}_n, \boldsymbol{\theta}\right)\big|\,\mathbf{x}_n, \boldsymbol{\theta}\right]\end{aligned} \tag{A.47}$$

where $f\left(a|\,\mathbf{x}_n, \boldsymbol{\theta}\right)$ is given by Equation A.24. Differentiating $\log f\left(A|\,\mathbf{x}_n, \boldsymbol{\theta}\right)$ with respect to θ_s yields

$$\begin{aligned}\frac{\partial}{\partial\theta_s}\log f\left(A|\mathbf{x}_n, \boldsymbol{\theta}\right) = {}& \frac{\partial}{\partial\theta_s}\log f\left(A|\,\boldsymbol{\theta}\right) - (N-n)\mathrm{E}\left[\frac{\partial}{\partial\theta_s}\log f\left(A|\,\boldsymbol{\theta}\right)|\mathbf{x}_n, \boldsymbol{\theta}\right]\\ &+(N-n-1)\mathrm{E}\left[\frac{\partial}{\partial\theta_s}\log \rho_f\left(\boldsymbol{\Lambda}|\,\boldsymbol{\theta}\right)|\,A, \mathbf{x}_n, \boldsymbol{\theta}\right]\end{aligned} \tag{A.48}$$

where $\boldsymbol{\Lambda}$, given $A = a$, has density $\eta\left(a|\lambda, \boldsymbol{\theta}\right)\xi\left(\lambda|\mathbf{x}_n, \boldsymbol{\theta}\right)/f\left(a|\mathbf{x}_n, \boldsymbol{\theta}\right)$. Now

$$\begin{aligned}&\mathrm{E}\left\{\frac{\partial}{\partial\theta_r}\log f\left(A|\boldsymbol{\theta}\right)\mathrm{E}\left[\frac{\partial}{\partial\theta_s}\log \rho_f\left(\boldsymbol{\Lambda}|\,\boldsymbol{\theta}\right)|\,A, \mathbf{x}_n, \boldsymbol{\theta}\right]|\,\mathbf{x}_n, \boldsymbol{\theta}\right\}\\ &= \int_0^\infty \mathrm{E}\left[\frac{\partial}{\partial\theta_r}\log f(A|\boldsymbol{\theta})|\,\lambda, \boldsymbol{\theta}\right]\mathrm{E}\left[\frac{\partial}{\partial\theta_s}\log f\left(A|\,\boldsymbol{\theta}\right)|\lambda, \boldsymbol{\theta}\right]\xi\left(\lambda|\mathbf{x}_n, \boldsymbol{\theta}\right)dx\end{aligned}$$

$$= \mathrm{Cov}\left\{ \mathrm{E}\left[\frac{\partial}{\partial \theta_r} \log f(A|\boldsymbol{\theta}) | \boldsymbol{\Lambda}, \boldsymbol{\theta} \right], \mathrm{E}\left[\frac{\partial}{\partial \theta_s} \log f(A|\boldsymbol{\theta}) | \boldsymbol{\Lambda}, \boldsymbol{\theta} \right] | \mathbf{x}_n, \boldsymbol{\theta} \right\}$$

$$+ \mathrm{E}\left[\frac{\partial}{\partial \theta_r} \log f(A|\boldsymbol{\theta}) | \mathbf{x}_n, \boldsymbol{\theta} \right] \mathrm{E}\left[\frac{\partial}{\partial \theta_s} \log f(A|\boldsymbol{\theta}) | \mathbf{x}_n, \boldsymbol{\theta} \right] \quad \text{(A.49)}$$

Combining Equations A.47, A.48, and A.49, we have

$$\frac{\partial^2}{\partial \theta_s \partial \theta_r} \log L = \mathrm{E}\left[\sum_{k=1}^{N} \frac{\partial^2}{\partial \theta_s \partial \theta_r} \log f(Y_k | \boldsymbol{\theta}) \mathbf{x}_n, \boldsymbol{\theta} \right]$$

$$+ (N-n)\,\mathrm{Cov}\left[\frac{\partial}{\partial \theta_r} \log f(A|\boldsymbol{\theta}), \frac{\partial}{\partial \theta_s} \log f(A | \mathbf{x}_n, \boldsymbol{\theta}) | \mathbf{x}_n, \boldsymbol{\theta} \right]$$

$$+ (N-n)(N-n-1)\,\mathrm{Cov}\left\{ \mathrm{E}\left[\frac{\partial}{\partial \theta_r} \log f(A|\boldsymbol{\theta}) | \boldsymbol{\Lambda}, \boldsymbol{\theta} \right], \right.$$

$$\left. \mathrm{E}\left[\frac{\partial}{\partial \theta_s} \log f(A|\boldsymbol{\theta}) | \boldsymbol{\Lambda}, \boldsymbol{\theta} \right] | \mathbf{x}_n, \boldsymbol{\theta} \right\} \quad \text{(A.50)}$$

But the last covariance term is equal to

$$\mathrm{Cov}\left[\frac{\partial}{\partial \theta_r} \log f(Y_{n+1} | \boldsymbol{\theta}), \frac{\partial}{\partial \theta_s} \log f(Y_{n+2} | \boldsymbol{\theta}) | \mathbf{x}_n, \boldsymbol{\theta} \right]$$

where the conditional joint density of (Y_{n+1}, Y_{n+2}) given $\mathbf{x}_n$ is obtained from Equation A.23 as

$$f(a_1, a_2 | \mathbf{x}_n, \boldsymbol{\theta}) = \int_0^{\infty} \eta(a_1 | \lambda, \boldsymbol{\theta}) \eta(a_2 | \lambda, \boldsymbol{\theta}) \varepsilon(\lambda | \mathbf{x}_n, \boldsymbol{\theta}) d\lambda \quad \text{(A.51)}$$

Therefore, the (r, s) entry of $\mathbf{I}_0(\boldsymbol{\theta})$ given by Equation A.21 is

$$\begin{aligned} \mathbf{I}_{0,rs}(\boldsymbol{\theta}) = {} & \mathrm{E}\left[\sum_{k=1}^{N} \frac{-\partial^2}{\partial \theta_s \partial \theta_r} \log f(Y_k | \boldsymbol{\theta}) | \mathbf{x}_n, \boldsymbol{\theta} \right] \\ & - \mathrm{Cov}\left[\sum_{k=1}^{N} \frac{\partial}{\partial \theta_r} \log f(Y_k | \boldsymbol{\theta}), \sum_{j=1}^{N} \frac{\partial}{\partial \theta_s} \log f(Y_j | \boldsymbol{\theta}) | \mathbf{x}_n, \boldsymbol{\theta} \right] \end{aligned} \quad \text{(A.52)}$$

The observed Fisher information matrix at $\boldsymbol{\theta}$ is the difference of the conditional expectation of the complete-data information matrix and the conditional covariance of the complete-data score functions, given the data $\mathbf{x}_n$. In the case of the regular exponential family in Equation A.45, the observed Fisher information matrix is

$$\mathbf{I}_0(\boldsymbol{\theta} | \mathbf{x}_n) = \mathrm{Cov}\left[t(\mathbf{Y}_N) | \boldsymbol{\theta} \right] - \mathrm{Cov}\left[t(\mathbf{Y}_N) | \mathbf{x}_n, \boldsymbol{\theta} \right] \quad \text{(A.53)}$$

where $t(\mathbf{Y}_N)=\sum_{k=1}^{N} t(Y_k)$, an $m\times 1$ vector of complete-data sufficient statistics. The formula in Equation A.53 was provided by Dempster et al. (1977) for the general incomplete-data problem. For our incomplete-data problem, explicit expressions for the covariances are given by Equations A.49 and A.50. These are in forms that can readily be computed.

The Fisher information matrix $\mathbf{I}(\boldsymbol{\theta})$ is the expectation of $-\partial^2/\partial\theta_r\partial\theta_s \log L$ with respect to the joint distribution of $X_1,\ldots,X_n$. From Equation A.52, this matrix is equal to the covariance matrix of the "estimated" complete-data score functions:

$$\mathrm{E}\left[\sum_{k=1}^{N} \frac{\partial}{\partial\theta_r} \log f\left(Y_k \,|\, \boldsymbol{\theta}\right) \Big| \, \mathbf{x}_n, \boldsymbol{\theta}\right], \quad r = 1, 2, \ldots, m$$

At the maximum-likelihood estimate $\hat{\theta}_n$, the Fisher information matrix $\mathbf{I}(\hat{\theta}_n)$ may be estimated by $\mathbf{I}_0(\hat{\theta}_n)$, with entries, evaluated at $\hat{\theta}_n$, that are given by Equation A.52; hence, an estimate of the asymptotic covariance matrix is given by the inverse of $\mathbf{I}_0(\hat{\theta}_n)$. Tests and confidence procedures can be obtained by the usual normal approximation.

Inference for $\boldsymbol{\theta}$ and *N*

In the previous section we demonstrated how maximum-likelihood estimates for $\boldsymbol{\theta}$ can be obtained when N is known. In this section, still assuming the weight function w(y) is given, we are interested in estimating both $\boldsymbol{\theta}$ and N. One approach is to do an $(m+1)$-dimensional grid search of a likelihood function $\mathrm{L}(\boldsymbol{\theta}, N \,|\, \mathbf{x}_n)$ based on Equation A.8. Another approach is to solve the likelihood equations in Equation A.18 via the EM algorithm to find $\hat{\boldsymbol{\theta}}(N)$ for different values of N, then determine $\hat{N}$ that maximizes the log-likelihood profile, $\log L\,(\hat{\boldsymbol{\theta}}(N), N|\mathbf{x}_n)$. The trouble with both of these approaches is that they are computationally expensive. In petroleum resource applications, our experience with the log-likelihood profile is that it is a rather "flat" function of N and frequently produces $\hat{N} = n$, and on occasion it produces an unacceptably large estimate of N.

A third approach is to ignore the superpopulation part completely and estimate N based on a method suggested by Gordon (1993) and then estimate $\boldsymbol{\theta}$ conditional on N. Gordon's idea is to split a successive sample from a finite population into two parts to approximate the unknown inclusion probabilities and then estimate N by an approximate Horvitz–Thompson-type estimator. His method requires solving a pair of transcendental equations that are symmetrical in form. Barouch et al. (1985) proposed an alternative pair that is asymmetrical and is competitive with Gordon's pair.

The fourth approach is to postulate that N also has a superpopulation probability function $\mathrm{P}(\bullet|\gamma)$ indexed by a vector of parameters $\boldsymbol{\gamma}$ and independent of the variate Y's, then derive an EM algorithm for both $\boldsymbol{\theta}$ and $\boldsymbol{\gamma}$. The probability function $\mathrm{P}(N|\boldsymbol{\gamma})$ may be interpreted as a model describing a random mechanism with regard to how N is generated or it may be considered as a prior distribution in an empirical Bayesian context. Here, the observations consist of $\mathbf{x}_n$ and $N \geq n$. The complete-data log likelihood is

$$\log L\left(y_1,\ldots,y_N,N\,|\,\boldsymbol{\theta},\boldsymbol{\gamma}\right)=\sum_{k=1}^{N}\log f\left(y_k\,|\,\boldsymbol{\theta}\right)+\log \mathrm{P}\left(N\,|\,\boldsymbol{\gamma}\right) \quad (A.54)$$

Let $(\boldsymbol{\theta}', \boldsymbol{\gamma}')$ denote the current estimate of $(\boldsymbol{\theta}, \boldsymbol{\gamma})$ and let $\mathbf{d} = \left\{\mathbf{x}_n, N \geq n\right\}$ denote the data. The M-step is to maximize over $(\boldsymbol{\theta}, \boldsymbol{\gamma})$ the following conditional expectation:

$$\begin{aligned} Q(\boldsymbol{\theta},\boldsymbol{\gamma}|\,\boldsymbol{\theta}',\boldsymbol{\gamma}') &= \mathrm{E}\left\{\log L\left(Y_1,\ldots,Y_N,\ N\,|\,\boldsymbol{\theta},\boldsymbol{\gamma}\right)|\,\mathbf{d},\boldsymbol{\theta}',\boldsymbol{\gamma}'\right\} \\ &= \mathrm{E}\left\{\mathrm{E}\left[\log L\left(Y_1,\ldots,Y_N,N\,|\,\boldsymbol{\theta},\boldsymbol{\gamma}\right)|\,N,\ \mathbf{x}_n,\boldsymbol{\theta}',\boldsymbol{\gamma}'\right]\mathbf{d},\ \boldsymbol{\theta}',\boldsymbol{\gamma}'\right\} \end{aligned} \quad (A.55)$$

Now, for $l = 0, 1, 2, \ldots,$

$$\begin{aligned} &\mathrm{E}\left[\log L\left(Y_1,\ldots,Y_N,N\,|\,\boldsymbol{\theta},\boldsymbol{\gamma}\right)|\,N=n+l,\ \mathbf{x}_n\ \boldsymbol{\theta}',\boldsymbol{\gamma}'\right] \\ &= \sum_{j=1}^{n}\log f\left(x_j\,|\,\boldsymbol{\theta}\right)+\log \mathrm{P}\left(n+l\,|\,\boldsymbol{\gamma}\right) \\ &\quad + l\,\mathrm{E}\left[\log f\left(Y_{n+1}\,|\,\boldsymbol{\theta}\right)|\ N=n+l,\ \mathbf{x}_n\ \boldsymbol{\theta}',\boldsymbol{\gamma}'\right] \end{aligned} \quad (A.56)$$

Therefore,

$$\begin{aligned} Q(\boldsymbol{\theta},\boldsymbol{\gamma}|\,\boldsymbol{\theta}',\boldsymbol{\gamma}') &= \sum_{j=1}^{n}\log f\left(x_j\,|\,\boldsymbol{\theta}\right)+\mathrm{E}\left[\log \mathrm{P}\left(N\,|\,\boldsymbol{\gamma}\right)|\,\mathbf{d},\ \boldsymbol{\theta}',\boldsymbol{\gamma}'\right] \\ &\quad + \mathrm{E}\left\{(N-n)\mathrm{E}\left[\log f\left(Y_{n+1}\,|\,\boldsymbol{\theta}\right)|\,N,\ \mathbf{x}_n\ \boldsymbol{\theta}',\boldsymbol{\gamma}'\right]|\,\mathbf{d},\boldsymbol{\theta}',\boldsymbol{\gamma}'\right\} \end{aligned} \quad (A.57)$$

The necessary conditions for $(\boldsymbol{\theta}, \boldsymbol{\gamma})$ to be a maximizer of $Q(\boldsymbol{\theta}, \boldsymbol{\gamma}|\boldsymbol{\theta}', \boldsymbol{\gamma}')$, are

$$\frac{\partial}{\partial \gamma_i}Q\left(\boldsymbol{\theta},\boldsymbol{\gamma}|\,\boldsymbol{\theta}',\boldsymbol{\gamma}'\right)=\mathrm{E}\left[\frac{\partial}{\partial \gamma_i}\log \mathrm{P}\left(N\,|\,\boldsymbol{\gamma}\right)|\,\mathbf{d},\ \boldsymbol{\theta}',\boldsymbol{\gamma}'\right]=0 \quad (A.58)$$

$$\frac{\partial}{\partial \gamma_i} Q(\boldsymbol{\theta}, \boldsymbol{\gamma} | \boldsymbol{\theta}', \boldsymbol{\gamma}') = \frac{n}{N_0} \bar{\mathrm{U}}_r(\boldsymbol{\theta} | \mathbf{x}_n)$$

$$+ \mathrm{E}\left\{\left(\frac{N-n}{N_0}\right) \mathrm{E}\left[\frac{\partial}{\partial \theta_r} \log f(Y_{n+1} | \boldsymbol{\theta}) | N, \mathbf{x}_n, \boldsymbol{\theta}', \boldsymbol{\gamma}'\right] | \mathbf{d}, \boldsymbol{\theta}', \boldsymbol{\gamma}'\right\} = 0$$

$i = 1, 2, ..., k \quad r = 1, 2, ..., m$

(A.59)

where $N_0 = \mathrm{E}(N|\mathbf{d}, \boldsymbol{\theta}', \boldsymbol{\gamma}')$. Note that when $\mathrm{P}(N|\boldsymbol{\gamma})$ is a point mass at $N_0 \geq n$, Equation A.59 reduces to Equation A.28.

We now derive the conditional expectations. Given for $(\boldsymbol{\theta}, \boldsymbol{\gamma})$, $\mathbf{x}_n$ and $N-n = l$ for $l = 0, 1, 2, \ldots$, define

$$q_l(\lambda | \boldsymbol{\theta}, \boldsymbol{\gamma}) = (n+l)!\, \rho_f(\lambda | \boldsymbol{\theta})^l\, \mathrm{P}(n+l | \boldsymbol{\gamma}) / l!, \qquad \lambda \geq 0 \qquad \text{(A.60)}$$

$$S_l(\boldsymbol{\theta}, \boldsymbol{\gamma} | \mathbf{x}_n) = \int_0^\infty q_l(\lambda | \boldsymbol{\theta}, \boldsymbol{\gamma})\, d\mathrm{G}_n(\lambda) \qquad \text{(A.61)}$$

$$s(\boldsymbol{\theta}, \boldsymbol{\gamma} | \mathbf{x}_n) = \sum_{l=0}^\infty S_l(\boldsymbol{\theta}, \boldsymbol{\gamma} | \mathbf{x}_n) \qquad \text{(A.62)}$$

Then the likelihood of $(\boldsymbol{\theta}, \boldsymbol{\gamma})$ given the data $\mathbf{d}$ is

$$L = \prod_{j=1}^n \left[\mathrm{w}(\mathrm{w}_j) f(x_j | \boldsymbol{\theta}) / \mathrm{b}_j\right] s(\boldsymbol{\theta}, \boldsymbol{\gamma} | \mathbf{x}_n) \qquad \text{(A.63)}$$

and the conditional probability function of N, given $\mathbf{d}$, is

$$\mathrm{P}(N = n+l | \mathbf{d}, \boldsymbol{\theta}, \boldsymbol{\gamma}) = S_l(\boldsymbol{\theta}, \boldsymbol{\gamma} | \mathbf{x}_n) / s(\boldsymbol{\theta}, \boldsymbol{\gamma} | \mathbf{x}_n) \qquad \text{(A.64)}$$

Now, for $\lambda \geq 0$ and $l \geq 0$, define a density similar to Equation A.19 as

$$\xi(\lambda | l, \mathbf{x}_n, \boldsymbol{\theta}, \boldsymbol{\gamma}) = q_l(\lambda | \boldsymbol{\theta}, \boldsymbol{\gamma})\, d\mathrm{G}_n(\lambda) / S_l(\boldsymbol{\theta}, \boldsymbol{\gamma} | \mathbf{x}_n) \qquad \text{(A.65)}$$

Let $\eta(a|\lambda, \boldsymbol{\theta})$ be the density given by Equation A.22. Then the conditional density of Y_{n+1}, given $N = n+1$ and $\mathbf{x}_n$ at $Y_{n+1} = a$, is

$$f(a | N = n+l, \mathbf{x}_n, \boldsymbol{\theta}, \boldsymbol{\gamma}) = \int_0^\infty \eta(a | \lambda, \boldsymbol{\theta})\, \xi(\lambda | l, \mathbf{x}_n, \boldsymbol{\theta}, \boldsymbol{\gamma})\, d\lambda \qquad \text{(A.66)}$$

Therefore the second term in the solution of Equation A.59 is equal to

$$\frac{1}{N_0}\sum_{l=1}^{\infty} l\,\mathrm{P}(N=n+l\,|\,\mathbf{d},\boldsymbol{\theta}',\boldsymbol{\gamma}')\int_0^{\infty}\frac{\partial}{\partial\theta_r}\log f(a\,|\,\boldsymbol{\theta})\int_0^{\infty}\eta(a\,|\,\lambda,\boldsymbol{\theta}')\xi(\lambda\,|\,l,\mathbf{x}_n,\boldsymbol{\theta}',\boldsymbol{\gamma}')\,d\lambda\,da$$
$$=\left(1-\frac{n}{N_0}\right)\int_0^{\infty}\mathrm{E}\left[\frac{\partial}{\partial\theta_r}\log f\,(A\,|\,\boldsymbol{\theta})\,|\,\lambda\,,\boldsymbol{\theta}'\right]\xi(\mathbf{d},\boldsymbol{\theta}',\boldsymbol{\gamma}')\,d\lambda \tag{A.67}$$

where $\xi(\lambda|\mathbf{d}, \boldsymbol{\theta}', \boldsymbol{\gamma}')$ is the mixture density function defined as

$$\xi(\lambda\,|\,\mathbf{d},\boldsymbol{\theta}',\boldsymbol{\gamma}')=\sum_{l=0}^{\infty}a_l\,\xi(l,\mathbf{x}_n,\boldsymbol{\theta}',\boldsymbol{\gamma}') \tag{A.68}$$

with

$$a_l = l\,\mathrm{P}(N=n+l\,|\,\mathbf{d},\boldsymbol{\theta}',\boldsymbol{\gamma}')/(N_0-n),\quad l=0,1,2,\ldots, \tag{A.69}$$

and

$$N_0 = n+\sum_{l=0}^{\infty} l\,\mathrm{P}(N=n+l\,|\,\mathbf{d},\boldsymbol{\theta}',\boldsymbol{\gamma}')$$
$$= n+l\sum_{l=0}^{\infty} l\,S_l(\boldsymbol{\theta}',\boldsymbol{\gamma}'\,|\,\mathbf{x}_n)/S(\boldsymbol{\theta}',\boldsymbol{\gamma}'\,|\,\mathbf{x}_n) \tag{A.70}$$

By defining $\boldsymbol{\Lambda}$ as a random variable with density given by Equation A.68, Equation A.59 is then reduced to

$$\frac{n}{N_0}\bar{\mathbf{U}}_r(\boldsymbol{\theta}\,|\,\mathbf{x}_n)+\left(1-\frac{n}{N_0}\right)\mathrm{E}\left[\frac{\partial}{\partial\theta_r}\log f\,(\boldsymbol{\Lambda}\,|\,\boldsymbol{\theta})\,|\,\mathbf{d},\boldsymbol{\theta}',\boldsymbol{\gamma}'\right]=0 \tag{A.71}$$

where the marginal distribution of A given data $\mathbf{d}$ has the same form as Equation A.24, except that ξ-density is given by Equation A.68. Comparing Equation A.28 and Equation A.71, we see that they have the same form.

To illustrate, let us assume that N is distributed according to a Poisson variate with mean $\boldsymbol{\gamma}$. Define

$$S_1(\boldsymbol{\theta},\boldsymbol{\gamma}\,|\,\mathbf{x}_n)=\int_0^{\infty}\exp\left\{-\boldsymbol{\gamma}\left[1-\rho_f(\lambda\,|\,\boldsymbol{\theta})\right]\right\}dG_n(\lambda)$$
$$S_\rho(\boldsymbol{\theta},\boldsymbol{\gamma}\,|\,\mathbf{x}_n)=\int_0^{\infty}\rho_f(\lambda\,|\,\boldsymbol{\theta})\exp\left\{-\boldsymbol{\gamma}\left[1-\rho_f(\lambda\,|\,\boldsymbol{\theta})\right]\right\}dG_n(\lambda) \tag{A.72}$$

Then $S_1(\boldsymbol{\theta}, \boldsymbol{\gamma} | \mathbf{x}_n) = \boldsymbol{\gamma}^n S_1(\boldsymbol{\theta}, \boldsymbol{\gamma} | \mathbf{x}_n)$, and from Equation A.64,

$$\mathrm{P}(N = n + l | \mathbf{d}, \boldsymbol{\theta}, \boldsymbol{\gamma}) = (\boldsymbol{\gamma}^l e^{-\gamma}/l!) \int_0^\infty \rho_f(\lambda | \boldsymbol{\theta})^l \, dG_n(\lambda) / S_1(\boldsymbol{\theta}, \boldsymbol{\gamma} | \mathbf{x}_n) \quad \text{(A.73)}$$

for $l = 0, 1, 2, \ldots$. Hence, $\mathrm{E}(N | \mathbf{d}, \boldsymbol{\theta}, \boldsymbol{\gamma}) = n + \boldsymbol{\gamma}\, S_\rho(\boldsymbol{\theta}, \boldsymbol{\gamma} | \mathbf{x}_n) / S_1(\boldsymbol{\theta}, \boldsymbol{\gamma} | \mathbf{x}_n)$ and the mixture density in Equation A.68 is given as

$$\varepsilon(\lambda | \mathbf{d}, \boldsymbol{\theta}, \boldsymbol{\gamma}) = \rho_f(\lambda | \boldsymbol{\theta}) \exp\{-\lambda[1 - \rho_f(\lambda | \boldsymbol{\theta})]\} \, dG_n(\lambda) / S_\rho(\boldsymbol{\theta}, \boldsymbol{\gamma} | \mathbf{x}_n) \quad \text{(A.74)}$$

The EM iteration $\left[\boldsymbol{\theta}^{(v)}, \boldsymbol{\gamma}^{(v)}\right] \to \left[\boldsymbol{\theta}^{(v+1)}, \boldsymbol{\gamma}^{(v+1)}\right]$ is given as

$$\boldsymbol{\gamma}^{(v+1)} = n + \boldsymbol{\gamma}^{(v)} S_\rho(\boldsymbol{\theta}^{(v)}, \boldsymbol{\gamma}^{(v)} | \mathbf{x}_n) / S_1(\boldsymbol{\theta}^{(v)}, \boldsymbol{\gamma}^{(v)} | \mathbf{x}_n) \quad \text{(A.75)}$$

and

$$\frac{n}{\boldsymbol{\gamma}^{(n+1)}} \bar{\mathbf{U}}_r(\boldsymbol{\theta}^{(v+1)} | \mathbf{x}_n) + \left(1 - \frac{n}{\boldsymbol{\gamma}^{(n+1)}}\right) \mathrm{E}\left[\frac{\partial}{\partial \theta_r} \log f(A | \boldsymbol{\theta}^{(v+1)}) \Big| \mathbf{d}, \boldsymbol{\theta}^{(v)}, \boldsymbol{\gamma}^{(v)}\right] = 0 \quad \text{(A.76)}$$

where $r = 1, 2, \ldots, m$. In the case of the lognormal superpopulation, Equation A.76 is reduced to Equation A.31 and Equation A.32 with N replaced by $\boldsymbol{\gamma}^{(v+1)}$. The conditional expectations of $\log A$ and $(\log A - \mu)^2$ given data $\mathbf{d}$ and $(\boldsymbol{\theta}', \boldsymbol{\gamma}')$ are given by Equations A.33 and A.34 with ξ-density replaced by that in Equation A.74. The EM iterations will always produce a pair of estimates $(\hat{\theta}_n, \hat{\gamma}_n)$.

From a computational point of view, the EM algorithm for the superpopulation approach to N is not any more difficult than for fixed N. The basic computation still lies in the accurate evaluation of the general gamma density $dG_n(\lambda)$.

This approach for predicting N is sufficiently general to include the usual Bayesian/subjective approach. In this case, the prior distribution can be any arbitrary but completely specified distribution, and it need not be a member of a parametric family. The posterior distribution of N given the data $\mathbf{d}$ is that given in Equation A.64 with $\mathrm{P}(\bullet|\gamma)$ replaced by the prior probability distribution. The EM algorithm for $\boldsymbol{\theta}$ solves Equation A.71 via this posterior distribution and $f(y|\boldsymbol{\theta})$. This approach to the estimation of N is attractive in petroleum resource evaluation because an explorationist quite frequently has other pertinent geological information about the number of fields/pools that could exist in the play. This information is usually summarized as a subjective distribution for N.

In a situation in which it is known that N has a finite support $N_1, ..., N_k$, but unknown probability masses $\gamma_1, ..., \gamma_k$ with $\sum \gamma_i = 1$, the maximum-likelihood estimation procedure for $(\boldsymbol{\theta}, \mathbf{r})$ is equivalent to the likelihood profile method on the support $N_1, ..., N_k$; hence, no advantage is gained. This equivalence can be seen as follows: For fixed $\boldsymbol{\theta}$, the likelihood is maximized by setting all γ_1's to zero except for the one γ_j with an associated N_j that maximizes the conditional likelihoods of $\boldsymbol{\theta}$ given $N = N_i$ for $i = 1, 2, ..., k$.

Inference for the Weight Function

In this section we shall consider the weight function $\mathrm{w}(y, \boldsymbol{\beta})$ where $\boldsymbol{\beta}$ is a vector of parameters $\beta_1, \ldots \beta_k$, and look at the joint maximum-likelihood estimation of $\boldsymbol{\theta}$ and $\boldsymbol{\beta}$. To this end, let $\rho_f(\lambda|\boldsymbol{\theta}, \boldsymbol{\beta})$ be the Laplace transform of $\mathrm{w}(y, \boldsymbol{\beta})$ with respect to $f(y|\boldsymbol{\theta})$, and let $d\mathrm{G}_n(\lambda|\boldsymbol{\beta})$ be the general gamma density with parameters $\mathrm{b}_j(\boldsymbol{\beta})$, where $j = 1, 2, \ldots, n$. Let $S(\boldsymbol{\theta}, \boldsymbol{\beta}|\mathbf{x}_n)$ be the integral given by Equation A.12 in terms of $\rho_f(\lambda|\boldsymbol{\theta}, \boldsymbol{\beta})$ and $d\mathrm{G}_n(\lambda|\boldsymbol{\beta})$. Therefore, the log likelihood of $(\boldsymbol{\theta}, \boldsymbol{\beta})$ is

$$\log L = \sum_{j=1}^{n} f(x_j|\boldsymbol{\theta}) + \sum_{j=1}^{n} \left[\mathrm{w}(x_j, \boldsymbol{\beta}) - \log \mathrm{b}_j(\boldsymbol{\beta}) \right] + \log S(\boldsymbol{\theta}, \boldsymbol{\beta}|\mathbf{x}_n) \quad \text{(A.77)}$$

We shall assume that $\mathrm{w}(y, \boldsymbol{\beta})$ is sufficiently smooth so that its partial derivatives with respect to β_l, $l = 1, 2, \ldots, k$, all exist and are continuous. Define

$$\dot{\mathrm{w}}_{lj}(\boldsymbol{\beta}) = \frac{\partial}{\partial \beta_l} \mathrm{w}(x_j, \boldsymbol{\beta})$$

and

$$A_{lj}(\boldsymbol{\beta}) = \sum_{i=j}^{n} \frac{\dot{\mathrm{w}}_{lj}(\boldsymbol{\beta})}{\mathrm{b}_j(\boldsymbol{\beta})}$$

$$l = 1, 2, \ldots, k; j = 1, 2, \ldots, n$$

Let $\varepsilon(\lambda|\mathbf{x}_n, \boldsymbol{\theta}, \boldsymbol{\beta})$ and $\eta(y|\boldsymbol{\theta}, \boldsymbol{\beta})$ be defined in the same manner as Equations A.19 and A.22 respectively. The score function with respect to β_l is therefore

$$\frac{\partial}{\partial \beta_l}\log L = \sum_{j=1}^{n}\left[\dot{\mathrm{w}}_{lj}(\boldsymbol{\beta}) - A_{lj}(\boldsymbol{\beta})\right]$$
$$+\int_0^{\infty}\left[(N-n)\frac{\partial}{\partial \beta_l}\log\varphi_f(\lambda|\boldsymbol{\theta},\boldsymbol{\beta}) + \frac{\partial}{\partial \beta_l}\log d\mathrm{G}_n(\lambda|\boldsymbol{\beta})\right]\xi(\lambda|\mathbf{x}_n,\boldsymbol{\theta},\boldsymbol{\beta})\,d\lambda$$
$$l = 1, 2, \ldots, k \qquad \text{(A.78)}$$

where

$$\frac{\partial}{\partial \beta_l}\log\varphi_f(\lambda|\boldsymbol{\theta},\boldsymbol{\beta}) = -\lambda \mathrm{E}\left[\dot{\mathrm{w}}(Y, \boldsymbol{\beta})|\lambda,\boldsymbol{\theta},\boldsymbol{\beta}\right] \qquad \text{(A.79)}$$

and $\frac{\partial}{\partial \beta_l}\log d\mathrm{G}_n(\lambda|\boldsymbol{\beta})$ can be approximated in the same way as $\log d\mathrm{G}_n(\lambda|\boldsymbol{\beta})$ (see Eqs. A.15 and A.17). The score function with respect to $\boldsymbol{\theta}$ is the same as Equation A.18. Maximum-likelihood estimates $(\hat{\boldsymbol{\theta}}, \hat{\boldsymbol{\beta}})$ are solutions to

$$\left(\frac{\partial}{\partial \theta_1}\log L, \ldots, \frac{\partial}{\partial \theta_m}\log L, \frac{\partial}{\partial \beta_1}\log L, \ldots, \frac{\partial}{\partial \beta_k}\log L\right) = \mathbf{0} \qquad \text{(A.80)}$$

A Newton–Raphson approach to the calculation of $(\hat{\boldsymbol{\theta}}, \hat{\boldsymbol{\beta}})$ would require second derivatives of log L and would be a mess—and computationally expensive.

When $\mathrm{w}(y, \boldsymbol{\beta})$ is parameterized by only one parameter, such as $\mathrm{w}(y, \boldsymbol{\beta}) = y^{\boldsymbol{\beta}}$, the maximum-likelihood estimates may be obtained by solving the equation

$$\frac{\partial}{\partial \theta_r}\log L = 0, \quad r = 1, 2, \ldots, m$$

by the EM algorithm to find $\hat{\boldsymbol{\theta}}(\boldsymbol{\beta})$ for each fixed $\boldsymbol{\beta}$. Then $\hat{\boldsymbol{\beta}}$ is determined from the log-likelihood profile $\log L\left[\hat{\boldsymbol{\theta}}(\boldsymbol{\beta}), \boldsymbol{\beta}\right]$ either graphically or via a one-dimensional, gradient-free maximization algorithm. The log-maximized relative-likelihood function

$$R_{\max}(\boldsymbol{\beta}) = \log L\left[\hat{\boldsymbol{\theta}}(\boldsymbol{\beta}), \boldsymbol{\beta}\right] - \log L(\hat{\boldsymbol{\theta}}, \hat{\boldsymbol{\beta}}) \qquad \text{(A.81)}$$

can be plotted to examine plausible values of $\boldsymbol{\beta}$. A large-sample approximation to the 95% confidence interval for $\boldsymbol{\beta}$ is obtainable from the likelihood ratio test as $\{\boldsymbol{\beta}: -2\, R_{\max}(\boldsymbol{\beta}) \leq 3.84\}$.

When $\boldsymbol{\beta}$ is of higher dimension, gradient-based procedures such as the quasi-Newton method or the conjugate gradient method can be used for maximizing log $L[\hat{\boldsymbol{\theta}}(\boldsymbol{\beta}), \boldsymbol{\beta}]$. As in the EM algorithm, these methods will also increase the log-likelihood profile at each iteration. In principle, the EM algorithm could also be used to find maximum-likelihood estimates for both $\boldsymbol{\theta}$ and $\boldsymbol{\beta}$ simultaneously. However, the part of the E-step that is relevant to $\boldsymbol{\beta}$ is difficult to carry out. It does not appear to have a sufficiently simple form for computations. We shall not give the expressions here.

Appendix B: Nonparametric Procedure for Estimating Distributions

Let F denote an underlying superpopulation distribution that is assumed to be discrete (i.e., that gives mass p_k to z_k, $k = 1, \dots, K$). In this appendix, we explain the procedure for estimating F without making any assumptions about its shape. Suppose there are N pools in a play with magnitudes (such as pool sizes) $Y_1, \dots, Y_N$. This model assumes that the N values are generated independently of an identical distribution, F.

Let $(y_1, \dots, y_n)$ denote the magnitudes of the n discovered pools, in order of discovery. Let N_k be the unknown number of Y_i's that have masses of z_k, and let n_k be the observed number of y_i's in the sample that have masses of z_k, $k = 1, \dots, K$. It is assumed that sampling is executed proportional to the size measure $\mathrm{w}(y)$.

Let $\mathrm{b}_i = \mathrm{w}(y_i) + \cdots + \mathrm{w}(y_n)$. It can then be shown that the probability of observing the ordered sample $(y_1, \dots, y_n)$ under the successive sampling discovery model is proportional to

$$L \propto \prod_{k=1}^{K} p_k^{n_k} \int_0^\infty \left[\sum_{k=1}^{K} p_k e^{-tw(z_k)} \right]^{N-n} g_n(t)\,dt \qquad \text{(B.1)}$$

where $g_n(t)$ is the density of $T = \varepsilon_1 / \mathrm{b}_1 + \cdots + \varepsilon_n / \mathrm{b}_n$ and the ε_i's are independent and identical standard exponential random variables (Wang and Nair, 1988). The nonparametric estimator can be obtained by maximizing this likelihood.

Under simple random sampling, the nonparametric estimator of F is given by the usual edf (empirical distribution function)

$$F_n(y) = \sum_{k:z_k \le y} \frac{n_k}{n} \tag{B.2}$$

This estimator is not valid here, however, because the sampling is biased. Thus the maximum-likelihood estimator of F for the successive sampling is now given by

$$\hat{F}_n(y) = \sum_{k:z_k \le y} \hat{p}_k \tag{B.3}$$

where $\{p_k\}_{k=1}^{K}$, with $\sum_{k=1}^{K} p_k = 1$, maximizes the log likelihood

$$\log L = \text{Constant} + \sum_{k=1}^{K} n_k \log p_k + \log \int_0^{\infty} \left[\sum_{k=1}^{K} p_k e^{-tw(z_k)} \right]^{N-n} g_n(t)\,dt \tag{B.4}$$

The values $\hat{p}_1, \ldots, \hat{p}_k$ are to be determined numerically so that the value of $\log L$ expressed by Equation B.4 is maximized. It can be shown that the maximized estimate for Equation B.4 is

$$\hat{p}_k^{(j+1)} = \frac{n}{N}\left(\frac{n_k}{N}\right) + \left(1 - \frac{n}{N}\right) \left\{ \frac{\displaystyle\int_0^{\infty} \frac{\hat{p}_k^{(j)} e^{-tw(z_k)}}{\sum_{l=1}^{K} \hat{p}_l\, e^{-tw(z_l)}} \left[\sum_{l=1}^{K} \hat{p}_l\, e^{-tw(z_l)} \right]^{N-n} g_n(t)\,dt}{\displaystyle\int_0^{\infty} \left[\sum_{l=1}^{K} \hat{p}_l\, e^{-tw(z_l)} \right]^{N-n} g_n(t)\,dt} \right\} \tag{B.5}$$

Note that the estimator is a convex combination of the usual estimator (the proportion of observed data in the k th cell) and a second term that is the expected proportion of the remaining (unobserved) observations from the k th cell. Several results follow from this estimator:

1. If $w(y)$ does not depend on y so that the sampling is, indeed, simple random sampling, then the estimator of F from Equation B.5 is reduced to Equation B.2.

2. If $n = N$, the estimator is also reduced to Equation B.2. This is because all the members have been observed from the finite population, and thus the sampling design itself is irrelevant. If all members of the finite population have been observed, then the best estimator of F is, of course, the usual edf estimator.
3. If $N \to \infty$ with fixed n, it can be shown that $\hat{p}_k \propto n_k / z_k$, the length-biased sampling estimator given by Cox (1969) (see Appendix A).
4. When $w(z_k)$ is large, the second term in Equation B.5 is small and so $\hat{p}_k = n_k / N$, implying that all the members in the finite population have, in fact, been discovered.

After $\hat{F}$ has been estimated, it is then considered to be the population distribution. Bootstrapped samples are randomly drawn from $\hat{F}$ to obtain a sample of size N. A sample of size n is simulated by the discovery process model with exploration efficiency β, which is also estimated from the nonparametric model. The μ and σ^2 are estimated from sample size n using the anchored method (Kaufman, 1986) (see Appendix A). These two sampling steps are repeated 5000 times. Standard deviations of μ and σ^2 are computed and their 95% intervals are then derived.

Appendix C: The Largest Pool Size and Its Distribution

The *r* th Largest Pool-Size Distribution

Let $X_1^*, X_2^*, \ldots, X_r^*$ be prospect potentials of a play and let $X_{(r)}^*$ be the *r*th largest prospect potential, $r = 1, 2, \ldots$; that is, $X_{(1)}^*$ is the largest prospect potential, $X_{(2)}^*$ is the second largest, and so on. Then the quantity

$$\mathrm{EPS}_r = \mathrm{E}\left[X_{(r)}^* \middle| X_{(r)}^* > 0\right] \tag{C.1}$$

is the expected size of the *r*th largest pool.

The distribution of $X_{(r)}^*$ has a discontinuous jump at zero. The probability mass at zero is given by

$$\begin{aligned}
\mathrm{P}\left[X_{(r)}^* = 0\right] &= \mathrm{P}(N \le r-1) \quad \text{for } x \ge 0 \\
\mathrm{P}\left[X_{(r)}^* > x\right] &= \sum_{n=r}^{\infty} \mathrm{P}\left[X_{(r)}^* > x \middle| N = n\right] \mathrm{P}(N = n) \\
&= \mathrm{P}\left[X_{(r)}^* > x, N \ge r\right] \qquad \text{(C.2)} \\
&= \sum_{n=r}^{\infty} \sum_{k=r}^{n} \mathrm{P}(\text{exactly } k \text{ pools have} \\
&\qquad \text{potential} > x \mid N = n)\, \mathrm{P}(N = n)
\end{aligned}$$

Because the probability that a pool has potential greater than x is $H(x)$,

$$\begin{aligned} &\text{P (exactly } k \text{ pools have potential} > x \mid N = n) \\ &= \binom{n}{k} H(x)^k [1 - H(x)]^{n-k} \end{aligned} \tag{C.3}$$

Therefore, the distribution of the rth largest pool is given by

$$\begin{aligned} L_r(x) &= \mathrm{P}\left[X^*_{(r)} > x \mid X^*_{(r)} > 0\right] \\ &= \sum_{n=r}^{\infty} \sum_{k=r}^{n} \binom{n}{k} H(x)^k \left[1 - H(x)\right]^{n-k} \frac{\mathrm{P}(N = n)}{\mathrm{P}(N \geq r)} \end{aligned} \tag{C.4}$$

for $x > 0$ and $r = 1, 2, \ldots$.

The density of the rth largest size is obtained by differentiating $l - L_r(x)$ with respect to x and is given by

$$l_r(x) = \sum_{n=r}^{\infty} r \binom{n}{r} H(x)^{r-l} \left[1 - H(x)\right]^{n-r} h(x) \frac{\mathrm{P}(N = n)}{\mathrm{P}(N \geq r)} \tag{C.5}$$

Therefore, the expected rth largest pool size is given by

$$\mathrm{EPS}_r = \sum_{n=r}^{\infty} r \binom{n}{r} \int_0^{\infty} x H(x)^{r-l} \left[1 - H(x)\right]^{n-r} h(x) dx \frac{\mathrm{P}(N = n)}{\mathrm{P}(N \geq r)} \tag{C.6}$$

By the definition of play resource, it must be true that the expected play resource equals the sum of

$$\mathrm{E}\left[X^*_{(r)}\right] = \mathrm{P}(N \geq r) \mathrm{E}\left[X^*_{(r)} \mid X^*_{(r)} > 0\right] \tag{C.7}$$

and

$$\begin{aligned} \sum_{r=l}^{\infty} \mathrm{E}\left[X^*_{(r)}\right] &= \sum_{n=l}^{\infty} n \mathrm{P}(N = n) \\ &\times \int_0^{\infty} \left[\sum_{r=l}^{\infty} \binom{n-l}{r-l} H(x)^{r-l} \left[1 - H(x)\right]^{n-r} \right] x h(x) dx \end{aligned} \tag{C.8}$$

By the binomial theorem, the expression inside the square brackets is l. Hence, by Equations C.8 and C.4, we have

$$\sum_{r=1}^{\infty} \mathrm{E}\left[X_{(r)}^{*}\right] = \sum_{n=1}^{\infty} n\mathrm{P}(N=n)\mathrm{E}[X] = \mathrm{E}[T] \tag{C.9}$$

Generation of Reservoir Parameters for a Given Pool Size

For the economic analysis of petroleum resources, it is necessary to find the conditional distribution $Z_1, Z_2, \ldots, Z_{p-1}$ of Equation C.3 for a given pool size x. This conditional distribution is also of interest in exploration.

In what follows, let us assume that $Z' = (Z_1, \ldots, Z_p)$ has a multivariate lognormal distribution with mean μ of dimension p and positive definite variance matrix $\mathbf{\Sigma}$. Let $Y_j = \ln Z_j$, for $j = 1, 2, \ldots, p$, and denote $Y^T = (Y_1, Y_2, \ldots, Y_p)$. Under the assumption of lognormality, the joint distribution of $\sum_{j=1}^{p} Y_j, Y_1, Y_2, \ldots, Y_{p-1}$ is multivariate normal, with mean

$$\mathbf{m}^T = \left(\sum_{j=1}^{p} \mu_j, m_{p-1} \right) \tag{C.10}$$

where $\mathbf{m}_{p-1}^T = (\mu_1, \ldots, \mu_{p-1})$ and variance matrix

$$\mathbf{V} = \begin{pmatrix} \sigma^2 & \mathbf{b}^T \\ \mathbf{b} & \mathbf{\Sigma}_{p-1} \end{pmatrix} \tag{C.11}$$

where

$$\sigma^2 = \mathrm{Var}\left(\sum_{j-1}^{p} Y_j \right) = a^T \mathbf{\Sigma} a$$

with $a^T = (1, 1, \ldots, 1)$ of dimension p,

$$\mathbf{b}^T = \left[\sum_{j=1}^{p} Y_j \operatorname{Cov}(Y_j, Y_1), \ldots, \sum_{j=1}^{p} Y_j \operatorname{Cov}(Y_j, Y_{p-1}) \right]$$

and $\mathbf{\Sigma}_{p-1}$ is the variance matrix of $(Y_1, \ldots, Y_{p-1})$. Hence it follows that the conditional distribution of $Y_1, \ldots, Y_{p-1}$, given that $\sum_{j=1}^{p} Y_j = \ln(x/c)$, is multivariate normal with mean

$$\mathbf{m}_{p-1}(x) = \mathbf{m}_{p-1} + \mathbf{b}\left(\frac{\ln(x/c) - \sum_{j=1}^{p} \mu_j}{\sigma^2}\right) \tag{C.12}$$

and variance matrix

$$\mathbf{V}_{p-1} = \mathbf{\Sigma}_{p-1} - \left(\mathbf{b}\,\mathbf{b}^T/\sigma^2\right) \tag{C.13}$$

Note that V_{p-1} is independent of the given pool size x.

Let

$$\mathbf{m}_{p-1}(x) = \begin{bmatrix} m_1(x) \\ \vdots \\ m_{p-1}(x) \end{bmatrix}$$

$$\mathbf{V}_{p-1} = \begin{bmatrix} v_{1,1} & \cdots & v_{1,p-1} \\ v_{p-1,1} & \cdots & v_{p-1,p-1} \end{bmatrix}$$

Then, for $i = 1, 2, ..., p-1$, we have

$$\mathrm{E}\left[Z_i \middle| X = x\right] = e^{m_i(x) + \frac{1}{2}v_{ii}} \tag{C.14}$$

and

$$\mathrm{Var}\left[Z_i \mid X = x\right] = e^{2m_i(x) + \frac{1}{2}v_{ii}(e^{ii} - 1)} \tag{C.15}$$

The conditional distribution of Z_p, given that $X = x$, is lognormal with parameters

$$\ln(x/c) - \sum_{j=1}^{p-1} m_j(x) \text{ and } a_{p-1}^T \mathbf{V}_{p-1}\, a_{p-1} \tag{C.16}$$

where $a_{p-1}^{T} = (1, 1, \ldots, 1)$ of dimension $p-1$. Therefore,

$$\mathrm{E}\left[Z_p \mid X = x\right] = \exp\left\{2\left[\ln(x/c) - \sum_{j=1}^{p-1} m_j(x)\right] + a_{p-1}^{T}\, \mathbf{V}_{p-1}\, a_{p-1}\right\} \quad (\mathrm{C}.17)$$

$$\mathrm{Var}\left[Z_p \mid X = x\right] = \exp\left\{2\left[\ln(x/c) - \sum_{j=1}^{p-1} m_j(x) + a_{p-1}^{T}\, \mathbf{V}_{p-1}\, a_{p-1}\right]\right\} \times \left[\exp\left(a_{p-1}^{T}\, \mathbf{V}_{p-1}\, a_{p-1}\right) - 1\right] \quad (\mathrm{C}.18)$$

Appendix D: Pool Size Conditional on Pool Ranks

Theorem 1

Let $X^*_{(r)}$ be the rth largest prospect potential of a play with a conditional pool-size distribution $H(x)$ and number-of-pools distribution $P(N=n)$. For $k \geq 1$, let $x_k < \cdots < x_1$ denote a sequence of known pool sizes and let $r_1 < r_2 \cdots < r_k$ denote the ranks among all pools, both discovered and undiscovered, of the given pool sizes. The conditional density of $X^*_{(r)}$, given that $X^*_{(r_1)} = x_1, \ldots, X^*_{(r_k)} = x_k$ and $X^*_{(r)} > 0$, denoted by $f(x|x_1, \ldots, x_k)$, is the following:

1. For $r_j < r < r_{j+1}$ and $x_{j+1} < x < x_j$,

$$f(x|x_1,\ldots,x_k) = \frac{\Gamma(r_{j+1} - r_j)}{\Gamma(r_{j+1} - r)\Gamma(r - r_j)} \times \frac{\left[H(x_{j+1}) - H(x)\right]^{r_{j+1}-r-1}\left[H(x) - H(x_j)\right]^{r-r_j-1}}{\left[H(x_{j+1}) - H(x_j)\right]^{r_{j+1}-r_j-r}} \mathrm{h}(x) \tag{D.1}$$

where $\Gamma(m) = (m-1)!$, $\mathrm{h}(x)$ is the probability density function of a pool size, and $H(x) = \int_x^\infty \mathrm{h}(z)\,dz = 1$ minus the cumulative distribution function.

2. For $r < r_1$ and $x_1 < x < \infty$,

$$f\left(x \middle| x_1, \ldots, x_k\right) = \frac{\Gamma(r_1)}{\Gamma(r_1 - r)\Gamma(r)} \times \frac{\left[H(x_1) - H(x)\right]^{r_1 - r - 1}\left[H(x)\right]^{r-1}}{\left[H\left(x_j\right)\right]^{r_1 - 1}} \mathrm{h}(x) \tag{D.2}$$

3. For $r > r_k$ and $0 < x < x_k$,

$$f\left(x \middle| x_1, \ldots, x_k\right) = C_r \sum_{n=r}^{\infty} \left\{ \frac{\Gamma(n+1)}{\Gamma(n-r+1)\Gamma(r-r_k)} \left[1 - H(x)\right]^{n-r} \times \left[H(x) - H(x_k)\right]^{r - r_k - 1} \mathrm{h}(x) \right\} \mathrm{P}(N = n) \tag{D.3}$$

where

$$C_r^{-1} = \sum_{n=r}^{\infty} \frac{\Gamma(n+1)}{\Gamma(n - r_k + 1)} \left[1 - H(x_k)\right]^{n - r_k} \mathrm{P}(N = n) \tag{D.4}$$

Note that the conditional distribution of the rth largest pool size for a given discovery record depends upon the record only through the most adjacent pool ranks and their sizes. Furthermore, in the preceding cases 1 and 2, the conditional pool size given a discovery record is independent of N, the number of pools in the play. For example, suppose the second largest pool has been discovered; then the size of the largest pool depends only upon the second largest pool size and the pool-size distribution $H(x)$, regardless of other discoveries and N.

Corollary

Let $\mathrm{EPS}_{r|\bullet}$ denote the conditional expectation of the rth largest pool size, given a discovery record. That is,

$$\mathrm{EPS}_{r|\bullet} = \mathrm{E}\left[X_{(r)}^* \middle| X_{(r_1)}^* = x_1, \ldots, X_{(r_k)}^* = x_k, X_{(r)}^* > 0 \right]$$

where the given discovery record is the collection $\{(r_i, x_i) : i = 1, \ldots, k\}$ of ranks and pool sizes satisfying the conditions in Theorem 1. Then

1. For $r_j < r < r_{j+1}$,

$$\mathrm{EPS}_{r|\bullet} = \frac{\Gamma(r_{j+1}-r_j)}{\Gamma(r_{j+1}-r)\,\Gamma(r-r_j)} \int_0^1 y(u|\,x_j, x_{j+1})\,(1-u)^{r_{j+1}-r-1}\, u^{r-r_j-1}\, du \quad \text{(D.5)}$$

where

$$y\left(u|\,x_j, x_{j+1}\right) = H^{-1}\left\{H(x_j) + u\left[\,H(x_{j+1}) - H(x_j)\right]\right\} \quad \text{(D.6)}$$

2. For $r < r_1$,

$$\mathrm{EPS}_{r|\bullet} = \frac{\Gamma(r_1)}{\Gamma(r_1 - r)\Gamma(r)} \int_0^1 y(u|\,x_1)(1-u)^{r_1-r-1}\, u^{r-1} du \quad \text{(D.7)}$$

where

$$y(u|\,x_1) = H^{-1}\left[u H(x_1)\right] \quad \text{(D.8)}$$

3. For $r > r_k$,

$$\begin{aligned} \mathrm{EPS}_{r|\bullet} = {} & C_r \sum_{n=r}^{\infty} \frac{\Gamma(n+1)}{\Gamma(n-r_k+1)} \left[1 - H(x_k)\right]^{n-r_k} \\ & \times \left[\frac{\Gamma(n-r_k+1)}{\Gamma(n-r+1)\Gamma(r-r_k)} \times \int_0^1 y(u|\,x_k)(1-u)^{n-r}\, u^{r-r_k-1} du\right] \mathrm{P}(N=n) \end{aligned} \quad \text{(D.9)}$$

where C_r is given by Equation D.4 and

$$y\left(u|\,x_k\right) = H^{-1}\left\{H(x_k) + u\left[1 - H(x_k)\right]\right\} \quad \text{(D.10)}$$

Theorem 2

Let ranks $r_1 < r_2$ be given. Let $f_{r_2}(x)$ denote the conditional density of $X^*_{(r2)}$, given that $X^*_{(r2)} > 0$. The probability density function of the ratio of pool sizes with the specified ranks, for $1 < w < \infty$, is given by

$$\begin{aligned} g(w) = {} & \int_0^{\infty} \frac{\Gamma(r_2)}{\Gamma(r_2 - r_1)\Gamma(r_1)} \\ & \times \frac{\left[H(x) - H(wx)\right]^{r_2-r_1-1} H(wx)^{r_1-1}}{H(x)^{r_2-1}} \times x h(wx) f_{r_2}(x) dx \end{aligned} \quad \text{(D.11)}$$

The expectation of the ratio is given by

$$\mathrm{ERPS}_{r_1,r_2} = \sum_{n=r_2}^{\infty} \frac{\Gamma(r_2)}{\Gamma(r_2-r_1)\Gamma(r_1)} \times \left[\int_0^1 J(u|n,r_2)(1-u)^{r_2-r_1-1}\,du\right]\frac{P(N=n)}{P(N\geq r_2)} \qquad \text{(D.12)}$$

where

$$J(u|n,r_2)=r_2\binom{n}{r_2}\int_0^{\infty}\frac{1}{x}H^{-1}\left[uH(x)\right]H(x)^{r_2-1}\left[1-H(x)\right]^{n-r_2}h(x)dx \qquad \text{(D.13)}$$

For proofs of Theorem 1 and its Corollary and Theorem 2, see Lee and Wang (1986) and refer to Appendix A.

References

Aitchison, J., and J. A. C. Brown. 1969. *The lognormal distribution with special reference to its uses in economics.* Dept. Applied Economics, monograph 5. Cambridge, UK: Cambridge Univ. Press.

Aitchison, J., and J. A. C. Brown. 1973. *The lognormal distribution with special reference to its uses in economics.* 4th ed. Dept. Applied Economics, monograph 5. Cambridge, UK: Cambridge Univ. Press.

Arps, J. J., and T. G. Roberts. 1958. Economics of drilling for Cretaceous oil on east flank of Denver–Julesburg basin. *Bull. Am. Assoc. Pet. Geologists* 42 (11):2549–2566.

Atwater, G. I. 1956. Future of Louisiana offshore oil province. *Bull. Am. Assoc. Pet. Geologists* 40 (11):2624–2634.

Baecher, G. B. 1979. Subjective sampling approaches to resource estimation. In *Methods and models for assessing energy resources,* ed. M. Grenon, 186–209. Oxford, UK: Pergamon Press.

Barclay, J. E., G. D. Holmstrom, P. J. Lee, R. I. Campbell, and G. E. Reinson. 1997. *Carboniferous and Permian gas resources in the Western Canada Sedimentary Basin, Interior Plains. Part I: Geological play analysis and resource assessment, 1–67.* Geol. Survey Canada Bull. 515. Ottawa, Canada: Natural Resources Canada.

Barouch, E., S. Chow, G. M. Kaufman, and T. H. Wright. 1985. Properties of successive sample moment estimators. *Studies in Applied Math.* 73 (3): 239–260.

Barouch, E., and G. M. Kaufman. 1976. Probabilistic modeling of oil and gas discovery. In *Energy—Mathematics and models,* ed. F. S. Roberts, 248–260. Philadelphia, PA: Soc. Industrial and Applied Math.

Barouch, E., and G. M. Kaufman. 1977. Estimation of undiscovered oil and gas. In *Mathematical aspects of production and distribution of energy,* ed. P. D. Lax, 77–91. Proc. Symp. in Applied Mathematics 21. Providence, RI: Am. Math. Soc.

Barouch, E., G. M. Kaufman, and J. Nelligan. 1983. Estimation of parameters of oil and gas discovery process models using the expectation–maximization algorithm. In *Energy modeling and simulation,* ed. A. S. Kydes et al., 109–117. Amsterdam, The Netherlands: North-Holland.

Bettini, C. 1987. *Forecasting population of undiscovered oil fields with the log-Pareto distribution.* Stanford Univ., Stanford, CA. PhD diss.

Bickel, P. J., and K. A. Doksum. 1977. *Mathematical statistics—Basic ideas and selected topics.* Oakland, CA: Holden-Day.

Bickel, P. J., V. N. Nair, and P. C. C. Wang. 1992. Nonparametric inference under biased sampling from a finite population: *Annals of Statistics* 20 (2):853–878.

Bird, T., J. E. Barclay, R. I. Campbell, and P. J. Lee. 1994. *Triassic gas resources of the Western Canada Sedimentary Basin, Interior Plains. Part I: Geological play analysis and resource assessment.* Geol. Survey Canada Bull. 483. Ottawa, Canada: Natural Resources Canada.

Bloomfield, P., K. S. Deffeyes, G. S. Watson, Y. Benjamini, and R. A. Stine. 1979. *Volume and area of oil fields and their impact on order of discovery: Resource Estimation and Validation Project.* Princeton, NJ: Statistics and Geology Depts., Princeton Univ.

BP. 2006 (June). *Quantifying energy—BP statistical review of world energy.* London, UK: BP.

Burrus, J., K. Osadetz, S. Wolf, B. Doligez, K. Visser, and D. Dearborn. 1996. A two-dimensional regional basin model of Williston Basin hydrocarbon system: *Bull. Am. Assoc. Pet. Geologists* 80 (2):265–291.

Cassel, C. M., C. E. Särndal, and J. H. Wretman. 1977. *Foundations of inference in survey sampling.* New York, NY: John Wiley.

Chambers, J. M., W. S. Cleveland, B. Kleiner, and P. A. Tukey. 1983. *Graphic methods for data analysis,* 191–242. Belmont, CA: Wadsworth International Group.

Chen, Zhuoheng. 1993. Quantification of petroleum resources through sampling from a parent field size distribution and as a function of basin yield. Norwegian Institute of Technology, Trondheim, Norway. PhD diss.

Chen, Zhuoheng, and K. G. Osadetz. 2006. Undiscovered petroleum accumulation mapping using model-based stochastic simulation. *Math. Geology* 38 (1):1–16.

Chen, Zhuoheng, and R. Sinding–Larsen. 1992. *Resource assessment using a modified anchored method.* Presented at the 29th International Geological Congress, Kyoto, Japan, August 24–September 4.

Cochran, W. G. 1939. The use of analysis of variance in enumeration by sampling. *J. Am. Stat. Assoc.* 34:492–510.

Coustau, H. 1981. Habitat of hydrocarbons and field size distribution—A first step towards ultimate reserve assessment. In *Assessment of undiscovered oil and gas,* ed. Committee for Coordination of Joint Prospecting for Mineral Resources in Asian Offshore Areas (CCOP), 180–194. CCOP tech. pub. 10. Bangkok, Thailand: United Nations ESCAP.

Coustau, H., P. J. Lee, J. Dupuy, and J. Junca. 1988. The resources of the Eastern Shetland Basin, North Sea: A comparison of evaluation methods. *Bull. Can. Pet. Geology* 36 (2):177–185.

Cox, D. R. 1969. Some sampling problems in technology. In *New developments in survey sampling,* ed. N. L. Johnson and H. Smith, Jr., 506–527. New York, NY: Wiley-Interscience.

Cox, D. R. 1972. Regression models and life-tables. *J. R. Stat. Soc. Ser. B* 34 (2):187–220.

Cox, D. R. 1975. Partial likelihood. *Biometrika* 62 (2):269–276.

Crump, K. S. 1976. Numerical inversion of Laplace transforms using a Fourier series approximation. *J. Assoc. Computing Machinery* 23 (1):89–96.

Davis, J. C., and T. Chang. 1989. Estimating potential for small fields in mature province. *Bull. Am. Assoc. Pet. Geologists* 73 (8):967–976.

Dempster, A. P., N. M. Laird, and D. B. Rubin. 1977. Maximum likelihood from incomplete data via the EM algorithm [with discussion]. *J. R. Stat. Soc. Ser. B* 39 (1):1–38.

Dolton, G. L. 1984. Basin assessment methods and approaches in the U.S. Geological Survey. In *Petroleum resource assessment,* ed. C. D. Masters, 4–23. Int. Union of Geol. Sci. Pub. 17. Paris: IUGS.

Drew, L. J. 1990. *Oil and gas forecasting: reflections of a petroleum geologist.* Studies in Mathematical Geology no. 2. New York, NY: Oxford University Press.

Drew, L. J., E. D. Attanasi, and J. R. Schuenemeyer. 1988. Observed oil and gas field size distributions: A consequence of the discovery process and prices of oil and gas. *Math. Geology* 20 (8):939–953.

Drew, L. J., and J. H. Schuenemeyer. 1993. The evolution and use of discovery process models at the U.S. Geological Survey. *Bull. Am. Assoc. Pet. Geologists* 77 (3):467–478.

Drew, L. J., J. H. Schuenemeyer, and W. J. Bawiec. 1982. *Estimation of the future rates of discovery of oil and gas discoveries in the Gulf of Mexico.* U.S. Geol. Survey prof. paper no. 1252. Washington, DC: U.S. Dept. Interior.

Drew, L. J., J. H. Schuenemeyer, and D. H. Root. 1980. *Petroleum resource appraisal and discovery rate forecasting in partially explored regions. Part A: An application to the Denver Basin.* U.S. Geol. Survey prof. paper no. 1138 A-C. Washington, DC: U.S. Dept. Interior.

du Rouchet, J. 1980. Le programme DIAGEN, deux procédures pour apprécier l'évolution chimique de la matière organique/The DIAGEN Program, two methods for calculating the diagenetic evolution of organic matter. *Bull. des Centres de Recherches Exploration-Production Elf-Aquitaine* 4 (2):813–831.

Energy, Mines and Resources Canada, 1977. *Oil and natural gas resource of Canada, 1976.* Report no. EP 77–1. Ottawa, Canada: EMRC.

Energy Resources Conservation Board. 1989. *Alberta's reserves of crude oil, gas, natural gas liquids, and sulphur.* Calgary, Canada: ERCB.

Forman, D. J., and A. L. Hinde. 1985. Improved statistical method for assessment of undiscovered petroleum resources. *Bull. Am. Assoc. Pet. Geologists* 69 (1):106–118.

Gao, Haiyu, Zhuoheng Chen, K. G. Osadetz, P. Hannigan, and C. Watson. 2000. A pool-based model of the spatial distribution of undiscovered petroleum resources. *Math. Geology* 32 (6):725–749.

Gehman, H. M., R. A. Baker, and D. A. White. 1981. Assessment methodology—An industry viewpoint. In *Assessment of undiscovered oil and gas,* ed. Committee for Coordination of Joint Prospecting for Mineral Resources in Asian Offshore Areas (CCOP), 113–121. CCOP tech. pub. 10. Bangkok, Thailand: United Nations ESCAP.

Gill, D. 1994. Niagaran reefs of northern Michigan. Part II: Resource appraisal. *J. Pet. Geology* 17:231–242.

Goff, J. C. 1983. Hydrocarbon generation and migration from Jurassic source rocks in the E Shetland Basin and Viking Graben of the northern North Sea. *J. Geol. Soc. London* 140 (3):445–474.

Gordon, L. 1983. Successive sampling in large finite populations. *Ann. Stat.* 11 (2):702–706.

Gordon, L. 1993. Estimation for large successive samples with unknown inclusion probabilities. *Adv. Appl. Math.* 14 (1):89–122.

Grenon, M. 1979. Introduction to Chapter 2. In *Methods and models for assessing energy resources. IIASA Conf. on Energy Resources,* ed. M. Grenon, 115–116. Oxford, UK: Pergamon Press.

Griffin, D. L. 1965a. *The Devonian Slave Point, Beaverhill Lake, and Muskwa formations of northeastern British Columbia and adjacent areas.* British Columbia Dept. Mines and Pet. Resources Bull. 50. Victoria, Canada: BC Dept. Mines and Pet. Resources.

Griffin, D. L. 1965b. The facies front of the Devonian Slave Point–Elk Point sequence in northeastern British Columbia and the Northwest Territories. *J. Can. Pet. Tech.* January–March:3–22.

Haun, J. D. 1975. *Methods of estimating the volume of undiscovered oil and gas resources.* Am. Assoc. Pet. Geologists Studies in geology no. 1. Tulsa, OK: AAPG.

Hemphill, C. R., R. I. Smith, and F. Szabo. 1968. *Geology of Beaverhill Lake reefs, Swan Hills area, Alberta,* 50–90. Am. Assoc. Pet. Geologists Memoir 14. Tulsa, OK: AAPG.

Houghton, J. C. 1988. Use of the truncated shifted Pareto distribution in assessing size distribution of oil and gas fields. *Math. Geology* 20 (8):907–937.

Johnson, N. L., and S. Kotz. 1970. *Distributions in statistics: Continuous univariate distributions.* Vol. 1. New York, NY: John Wiley.

Kalbfleisch, J. D., and R. L. Prentice. 1973. Marginal likelihoods based on Cox's regression and life model. *Biometrika* 60 (2):267–278.

Kaufman, G. M. 1963. *Statistical decision and related techniques in oil and gas exploration.* Englewood Cliffs, NJ: Prentice-Hall.

Kaufman, G. M. 1965. Statistical analysis of the size distribution of oil and gas fields. In *Soc. Pet. Eng. of AIME, Symp. on Petroleum Economics and Evaluation,* 109–124. SPE preprint no. 1096. Richardson, TX: SPE.

Kaufman, G. M. 1986. Finite population sampling methods for oil and gas resource estimation. In *Oil and gas assessment—Methods and applications,* ed. D. D. Rice, 43–53. Am. Assoc. Pet. Geologists Studies in geology no. 21. Tulsa, OK: AAPG.

Kaufman, G. M., Y. Balcer, and D. Kruyt. 1975. A probabilistic model of oil and gas discovery. In *Methods of estimating the volume of undiscovered oil and gas resources,* ed. J. D. Haun, 113–142. Am. Assoc. Pet. Geologists Studies in geology no. 1. Tulsa, OK: AAPG.

Kingston, D. R., C. P. Dishroon, and P. A. Williams. 1983a. Global basin classification system. *Bull. Am. Assoc. Pet. Geologists* 67 (12):2175–2193.

Kingston, D. R., C. P. Dishroon, and P. A. Williams. 1983b. Hydrocarbon plays and global basin classification. *Bull. Am. Assoc. Pet. Geologists* 67 (12):2194–2198.

Kingston, D. R., C. P. Dishroon, and P. A. Williams. 1985. A hydrocarbon exploration crustal classification. *Oil Gas J.* 83 (10):146–156.

Klemme, H. D. 1975. Giant oil fields related to their geologic setting—A possible guide to exploration. *Bull. Can. Pet. Geology* 23 (1):30–66.

Klemme, H. D. 1986. Field size distribution related to basin characteristics. In *Oil and gas assessment—Methods and applications,* ed. D. D. Rice, 85–99. Am. Assoc. Pet. Geologists Studies in geology no. 21. Tulsa, OK: AAPG.

Lee, F. C., and P. J. Lee. 1994. *Petroleum resource assessments—A fractal approach,* 265–270. Geol. Survey Canada Current research 1994–E. Ottawa, Canada: Natural Resources Canada.

Lee, P. J. 1993a. *Lognormal and nonparametric discovery process models: Reliable resource assessment tools?* Presented at the Am. Assoc. Pet. Geologists annual convention, New Orleans, LA, April 24–26.

Lee, P. J. 1993b. *Oil and gas pool size probability distributions: J-shaped, lognormal, or Pareto?,* 93–96. Geol. Survey Canada Current research 1993–E. Ottawa, Canada: Natural Resources Canada.

Lee, P. J. 1993c. *The GSC lognormal discovery process model revisited.* Presented at the Int. Assoc. Math. Geology silver anniversary meeting—IAMG'93, Prague, Czechoslovakia, October 10–15.

Lee, P. J. 1993d. Two decades of petroleum resource assessments in the Geological Survey of Canada. *Can. J. Earth Sci.* 30:321–332.

Lee, P. J. 1997. Estimating number-of-pools distribution based on discovery sequence. In *Proceedings of the Third Annual Conference of the International Association for Mathematical Geology,* V. Pawlowsky–Glahn, 519. Vol. 2. Barcelona, Spain: CIMNE.

Lee, P. J. 1998. Analyzing multivariate oil and gas discovery data. In *Proceedings of the Fourth Annual Conference of the International Association for Mathematical Geology,* ed. A. Buccianti, G. Nardi, and R. Potenza, 451–456. Vol. 1. Naples, Italy: De Frede Editore.

Lee, P. J., J. Y. Chang, and H. P. Tzeng. 1999. *Petroleum Resources Information Management Evaluation System PETRIMES/W working guide.* Report no. NCS–87–CPC–M–006–001. National Cheng Kung Univ., Tainan, Taiwan.

Lee, P. J., and D. Gill. 1999. Comparison of discovery process methods for estimating undiscovered resources. *Bull. Can. Pet. Geology* 47 (1):19–30.

Lee, P. J., K. Olsen–Heise, and H. P. Tzeng. 1995. *Contrasts between the GSC and CPGC procedures for estimating undiscovered gas resources,* 337–340. Geol. Survey Canada Open file no. 3058. Ottawa, Canada: Natural Resources Canada.

Lee, P. J., and P. R. Price. 1991. Successes in 1980s bode well for W. Canada search. *Oil Gas J.* 89:92–97.

Lee, P. J., Ruo-Zhe Qin, and Yan-Min Shi. 1989. Conditional probability analysis of geological risk factors. In *Statistical applications in the earth sciences,* ed. F. P. Agterberg and G. F. Bonham–Carter, 271–276. Geol. Survey Canada Paper no. 89–9. Ottawa, Canada: Natural Resources Canada.

Lee, P. J., and D. A. Singer. 1994. Using PETRIMES to estimate mercury deposits in California. *Nonrenewable Resources* 3 (3):190–199.

Lee, P. J., and H. P. Tzeng. 1993. *The petroleum exploration and resource evaluation system (PETRIMES)—Working reference guide.* Version 3.0 (HP and PC). Geol. Survey Canada Open file no. 2703. Calgary: GSC Inst. Sedimentary and Pet. Geology.

Lee, P. J., and P. C. C. Wang. 1983a. Conditional analysis for petroleum resource evaluations. *Math. Geology* 15 (2):353–365.

Lee, P. J., and P. C. C. Wang. 1983b. Probabilistic formulation of a method for the evaluation of petroleum resources. *Math. Geology* 15 (1):163–181.

Lee, P. J., and P. C. C. Wang. 1984. PRIMES—A petroleum resources information management and evaluation system. *Oil Gas J.* 82 (40):204–206.

Lee, P. J., and P. C. C. Wang. 1985. Prediction of oil or gas pool sizes when discovery record is available. *Math. Geology* 17 (2):95–113.

Lee, P. J., and P. C. C. Wang. 1986. Evaluation of petroleum resources from pool size distribution. In *Oil and gas assessment—Methods and applications,* ed. D. D. Rice, 33–42. Am. Assoc. Pet. Geologists Studies in geology no. 21. Tulsa, OK: AAPG.

Lee, P. J., and P. C. C. Wang. 1987. *Petroleum resource evaluation concepts.* Presented at the Int. Union of Geol. Sci. Loen Conf., Loen, Norway, Sept. 29–Oct. 2.

Lee, P. J., and P. C. C. Wang. 1990. *An introduction to petroleum resource evaluation methods, CPSG Short Course SC-2.* Presented at the Canadian Soc. Pet. Geologists Convention, Calgary, May 27–30, Geol. Survey Canada contrib. no. 51789.

Lepoutré, M. 1986. DIAGEN—A numerical model for appreciation of chemical evolution of organic matter during time. In *Thermal modeling in sedimentary basins,* ed. J. Burrus, 247–256. Houston, TX: Gulf Publ.

Long, K. R. 1988. *Estimating the number and sizes of undiscovered oil and gas pools.* Univ. Arizona, Tucson, AZ. PhD diss.

Lorentziadis, P. L. 1991. *Forecasts in oil exploration and prospect evaluation for financial decisions: A semi-parametric approach.* Univ. California, Berkeley, CA. PhD diss.

Mast, R. F., G. L. Dolton, R. A. Crovelli, D. H. Root, E. D. Attanasi, P. E. Martin, L. W. Cooke, G. B. Carpenter, W. C. Pecora, and M. B. Rose. 1989. *Estimates of undiscovered conventional oil and gas resources in the United States—A part of the nation's energy endowment.* U.S. Geol. Survey and Minerals Mgmt. Service. Washington, DC: U.S. Dept. Interior.

Masters, C. D. 1985. *Petroleum resource assessment.* Int. Union of Geol. Sci. Pub. 17. Paris, France: IUGS.

McCrossan, R. G. 1969. An analysis of size frequency distribution of oil and gas reserves of western Canada. *Can. J. Earth Sci.,* 6 (2):201–211.

Meisner, J., and F. Demirmen. 1981. The creaming method—A Bayesian procedure to forecast future oil and gas discoveries in mature exploration provinces. *J. Roy. Stat. Soc.*, Ser. A, 144 (part I):1–13.

Neyman, J., and E. L. Scott. 1971. Outlier proneness of phenomena and of related distributions. In *Optimizing methods in statistics,* ed. J. S. Rustagi, 413–430. New York, NY: Academic Press.

Osadetz, K. G., P. J. Lee, P. K. Hannigan, and K. Olsen–Heise. 1995. *Natural gas resources of foreland belt of the Cordilleran orogen in Canada,* 345–348. Geol. Survey Canada Open file no. 3058. Ottawa, Canada: Natural Resources Canada.

Patil, G. P., and C. R. Rao. 1977. The weighted distributions—A survey and their applications. In *Application of statistics,* ed. P. R. Krishnaiah, 383–405. Amsterdam, The Netherlands: North-Holland.

Patil, G .P., and C. R. Rao. 1978. Weighted distributions and size-biased sampling with applications to wildlife populations and human families. *Biometrics* 34 (2):179–189.

Podruski, J. A., J. E. Barclay, A. P. Hamblin, P. J. Lee, K. G. Osadetz, R. M. Procter, and G. C. Taylor. 1988. *Conventional oil resources of western Canada. Part I. Resource endowment,* 1–125. Geol. Survey Canada Paper no. 87–26. Ottawa, Canada: Natural Resources Canada.

Power, M. 1992. Lognormality in the observed size distribution of oil and gas as a consequence of sampling bias. *Math. Geology* 24 (8):929–946.

Reinson, G. E., P. J. Lee, W. Warters, K. G. Osadetz, L. L. Bell, P. R. Price, F. Trollope, R. I. Campbell, and J. E. Barclay. 1993. *Devonian gas resources of Western Canada Sedimentary Basin—Play definition and resource assessment.* Geol. Survey Canada Bull. 452. Ottawa, Canada: Natural Resources Canada.

Rice, D. D. 1986. *Oil and gas assessment—Methods and applications.* Am. Assoc. Pet. Geologists Studies in geology no. 21. Tulsa, OK: AAPG.

Roy, K. J. 1979. Hydrocarbon assessment using subjective probability and Monte Carlo methods. In *Methods and models for assessing energy resources,* ed. M. Grenon, 279–290. Oxford, UK: Pergamon Press.

Schuenemeyer, J. H., and L. J. Drew. 1983. A procedure to estimate the parent population of the size of oil and gas fields as revealed by a study of economic truncation. *Math. Geology* 15 (1):145–162.

Smith, J. L., and G. L. Ward. 1981. Maximum likelihood estimates of the size distribution of North Sea oil fields. *Math. Geology* 13 (5):399–413.

Vardi, Y. 1982. Nonparametric estimation in the presence of length bias. *Ann. Stat.* 10 (2):616–620.

Vardi, Y. 1985. Empirical distributions in selection bias models. *Ann. Stat.* 13 (1):178–203.

Velleman, P. F., and D. C. Hoaglin. 1981. *Applications, basics, and computing of exploratory data analysis.* Boston, MA: Duxbury Press Div., Wadsworth.

Wang, P. C. C., and V. N. Nair. 1988. Statistical analysis of oil and gas discovery data. In *Quantitative analysis of mineral and energy resources,* ed. C. F. Chung, A. G. Fabbri, and R. Sinding–Larsen, 199–214. Dordrecht, The Netherlands: D. Reidel Publ.

White, D. A. 1980. Assessing oil and gas plays in facies-cycle wedge. *Bull. Am. Assoc. Pet. Geologists* 64 (8):1158–1178.

White, D. A., and H. M. Gehman. 1979. Methods of estimating oil and gas resources. *Bull. Am. Assoc. Pet. Geologists* 63 (12):2183–2192.

Williams, G. K. 1984. Some musings on the Devonian Elk Point Basin, western Canada. *Bull. Can. Pet. Geology* 32 (2):216–232.

Wilson, J. L., and C. Jordan. 1983. Middle shelf environment. In *Carbonate depositional environments,* ed. P. A. Scholle, D. G. Bebout, and C. H. Moore, 297–343 (335; Fig. 64). Am. Assoc. Pet. Geologists Memoir 33. Tulsa, OK: AAPG.

Wu, C. F. J. 1983. On the convergence properties of the EM algorithm. *Ann. Stat.* 11 (1):95–103.

Zipf, G. K. 1949. *Human behaviour and the principle of least effort.* Cambridge, MA: Addison-Wesley.

Index